Asheville-Buncombe
Technical Community College
Learning Resources Center
340 Victoria Road
Asheville, NC 28801

DISCARDED

AUG 4 2025

D1768896

Religion and Ecology in India and Southeast Asia

'David Gosling is uniquely placed to write this book. He is a trained and qualified scientist who then immersed himself in the religions and cultures of India and southeast Asia. He has "hands-on" experience in drawing science and religion together in facing practical issues of ecology.'

Professor John Bowker, Fellow of Gresham College, London

The resolution of the ecological problems facing the Indian sub-continent – with its huge Himalayan land mass and large population – and southeast Asia will be a major factor in whether life as we know it will survive beyond the early centuries of the new millennium. *Religion and Ecology in India and Southeast Asia* looks at the part the Hindu and Buddhist traditions could play in promoting more just and sustainable relationships between people and the natural world.

The ecological potential of these traditions is considered from both a historical perspective, and in relation to their contemporary expressions. From this standpoint, transformations between the past and the present are shown to offer the most fertile possibilities for improvement. Thus India's sacred groves, the conservation of medicinal herbs by Thai monks compensating for biodiversity loss, Hindu scientists acknowledging Vedānta in exploring the possibility of pain in plants, and Thai lay nuns redefining their roles in line with tradition and modernity, illustrate ways in which the Hindu and Buddhist traditions can improve ecological relationships.

David Gosling's arguments are based on his extensive fieldwork in the region and are framed by the socio-political context of religious change in India and southeast Asia, where it is maintained that the development-led analysis of Amartya Sen, with its emphasis on participative education, healthcare and a reduction of the gender imbalance, is a crucial prerequisite for social and environmental improvement. Though regional in scope, this is a study of global significance, considering the most urgent social and environmental problems of the new millennium.

David L. Gosling trained as a nuclear physicist and more recently was the first Spalding Fellow in Religions at Clare Hall, University of Cambridge, where he is currently based. He was Director of Church and Society at the World Council of Churches in Geneva and has published extensively on environmental issues in south Asia.

Religion and Ecology in India and Southeast Asia

David L. Gosling
with a foreword by Ninian Smart

London and New York

First published 2001
by Routledge
11 New Fetter Lane, London EC4P 4EE

Simultaneously published in the USA and Canada
by Routledge

29 West 35th Street, New York, NY 10001

Routledge is an imprint of the Taylor & Francis Group

© 2001 David L. Gosling

Typeset in Times by
M Rules
Printed and bound in Great Britain by
Biddles Ltd, Guildford and King's Lynn

All rights reserved. No part of this book may be reprinted or
reproduced or utilized in any form or by any electronic,
mechanical, or other means, now known or hereafter
invented, including photocopying and recording, or in any
information storage or retrieval system, without permission in
writing from the publishers.

British Library Cataloguing in Publication Data
A catalogue record for this book is available from the British Library

Library of Congress Cataloguing in Publication Data
A catalog record for this book has been requested

ISBN 0-415-24030-1 (hbk)
ISBN 0-415-24031-X (pbk)

This book is dedicated to all my friends in India and southeast Asia, in gratitude for their wisdom, inspiration and affection.

> The world is too much with us; late and soon,
> Getting and spending, we lay waste our powers:
> Little we see in Nature that is ours;
> We have given our hearts away, a sordid boon!
> The sea that bares her bosom to the moon;
> The winds that will be howling at all hours,
> And are up-gather'd now like sleeping flowers;
> For this, for everything, we are out of tune . . .
>
> William Wordsworth, from
Miscellaneous Sonnets, no. 23

Contents

List of figures x
Foreword xi
Acknowledgements xii

1 **Introduction** 1
  The Asian viewpoint 2
  Ecology 4
  Religion 6
  Methodology 8
  Main arguments 10
  Religion at Rio 13

2 **Ecology and Hindu tradition** 16
  Modes and habitats 17
  Nomads and settlers 18
  The early Vedic period 21
  The Upanishads 24
  The post-Vedic period 27
  A contemporary approach 29

3 **Ecology and modern India** 34
  Extracting resources 35
  The Reformers 36
  The impact of science 40
  Indian scientists 42
  Jagadish Chandra Bose 43
  Gandhian ethics 45

4 **Struggles for the forests** 51
  The Himalayas 52
  Colonial forestry 54
  Early opposition 56

	Chipko and Appiko	58
	Sunderlal Bahuguna	61
	Anna Hazare	63
5	**Ecology and Buddhism**	**68**
	Early Buddhism	69
	The Mahāyāna	72
	The spread of Buddhism	74
	Southeast Asia	76
	Ladakh	78
	Bhutan	82
6	**Thailand: a case study**	**86**
	Environmental parameters	88
	Ashoka's legacy	89
	Monastic reforms	91
	Buddhadāsa Bhikkhu	92
	Social and environmental activities	96
	Urban sects and movements	100
	The culture of gender	103
7	**India since Independence**	**110**
	The secular state	111
	'Abolish poverty'	113
	Patterns of resource use	116
	Environmental politics	118
	The Hindu Right	122
	Restructuring society	126
	Beyond liberalization	131
8	**Signs of hope**	**136**
	Global-level initiatives	137
	Personal initiatives	140
	North India	143
	South India	145
	Points of view	148
	Karan Singh	148
	M. C. Mehta	150
	Amartya Sen	153
	Sacred groves	155
9	**Expanding our horizons**	**159**
	Priorities and perceptions	159
	Religious misconceptions	160
	Continuities	162

Transformations	165
Public-interest science	169
Environmentalists at the crossroads	171
Secular India	173
Appendix A: Medicinal plants identified in Thailand	**176**
Appendix B: Indian non-governmental organizations	**181**
Select glossary	188
Notes	190
Select bibliography	201
Index	205

Figures

1.1	India's states	15
6.1	Thailand: fieldwork locations	87
6.2	The appropriateness of social roles for *mae chii*	106
7.1	Patterns of resource use in India	119

Foreword

We live in a time of globalization. And a vital and highly influential part of the world is to be found in India and southeast Asia. It is because of their populations, resources and conceptual traditions that they need a shrewd discussion, and here they get it. David Gosling is highly qualified, in both physics and the study of religion, to address the issues of ecology and the relevance to them of the Hindu and Buddhist traditions. Southern Asia is destined to play a crucial part in the preservation of our fragile planet.

India is particularly important to the future of our world. Its vast population (one in six humans lives there) and its simultaneous prosperity and great poverty put strains upon its traditional relationship to the environment. Nevertheless, importantly, the Hindu and Buddhist traditions can help us to see that our life is inextricably bound up with the natural world and the life of animals. Whether we believe in reincarnation or not, it illustrates an attitude which is important. One of the striking features of both traditions is their simultaneous concern with science and spirituality. India's great heritage of *yoga* and devotionalism is complemented with its investment in high technology.

The range of Dr Gosling's treatment of the issues is considerable. He presents not only the traditional roots involved, but also case studies in various modern areas. His summation of the varying challenges relating to southern Asian development and ecology rejects any simplistic attitudes and questions the opinions of some western environmentalists. His earlier researches on India and Thailand are brought to bear on contemporary Asian attitudes to science and religion. He illuminates the roots in Asia of a sane environmental system of values. He depicts with clarity the theories and philosophies of Amartya Sen, Vivekananda and Gandhi. Consequently, his contribution to the contemporary debate – both supporting scientific theories and also criticizing some aspects of contemporary global capitalism, not to mention outmoded social attitudes – shows that our planet is too important to be sacrificed to greed, desperation and ignorance. Dr Gosling is a wise guide to the field and an excellent antidote to these diseases among humankind.

Ninian Smart, J. F. Rowny Professor of Religious Studies
University College of Santa Barbara
California

Acknowledgements

I should like to thank the Spalding Trustees and the President and members of Clare Hall, University of Cambridge, for the Spalding Fellowship in 1992 which made possible the research on which much of this book is based. I am especially grateful to Julius Lipner, Michael Loewe and Carmen Blacker for their encouragement. I also thank John Sweet and the Cambridge Committee (CCCWD) which he chairs, and the USPG for further financial assistance and for appointing me as Cambridge Teape Fellow in Delhi.

In India I am grateful to the Principal and members of St Stephen's College, University of Delhi, for giving me the opportunity to teach a new course in environmental chemistry. Prem Sagar Dwivedi, formerly of St Stephen's, was especially helpful in translating and commenting on various texts. I wish to thank Kabir Mustafi and his colleagues at Bishop Cotton School, Shimla, and Mrinal Miri and the faculty members of the Indian Institute of Advanced Studies, also in Shimla, for their assistance.

In Thailand I should like to thank William Klausner, Sulak Sivaraksa, Chai-anan Samudavanija, Chatsumarn Kabilsingh and Phichai Tovivich for their support.

I am extremely grateful to Amartya K. Sen, Arvind N. Das, John D. Smith, Chris Hope, Peter Harvey and Gavin Flood for reading sections of the text. Ghillean Prance very kindly went over the botanical list in Appendix A.

I am enormously grateful to Rosemary Smith for her painstaking work in typing the manuscript.

Some of the earlier Thai fieldwork was supported by grants from the British Academy, the Nuffield Foundation and both the US and UK Social Science Research Councils. An initial grant by the UK Social Science Research Council was criticized in the House of Lords as a waste of public money (*Hansard*, 23 March 1976), and although the government spokesperson defended the award, subsequent funds were not forthcoming. However, private funding agencies have continued to support the research, all of which has been published (see Select bibliography p. 201).

David L. Gosling
University of Cambridge
Clare Hall, Cambridge

1 Introduction

At the start of another millennium the physical condition of our planet continues to deteriorate. From the standpoint of the industrial world, standards of living are rising for most, and patterns of employment are increasingly shifting from goods to services, with consequent reductions in growth rates and environmental impact. Some of the largest developing societies, however, experience severe poverty and the escalation of environmental problems such as deforestation, the pollution of waterways and the depletion of natural resources such as minerals, fuels and biodiversity.

People tend to view these issues according to where they live. From the perspective of the industrial nations, the world's most serious environmental problems are mainly the anticipated effects of global warming, ozone layer depletion and high population growth rates elsewhere. In the developing world, however, the perceived environmental problems cannot be so easily separated from their social context, but tend to be associated with loss of forest cover (and its inappropriate replacement), biodiversity loss, the contamination of rivers and natural resource depletion. Desertification is a major problem in some areas. Lesser problems are perceived to be urban traffic pollution, overfishing and the destruction of marine environments, and the environmental impact of large populations.

History and economics lay bare the patterns of exploitation and dependency that exist between industrial and developing nations. These range from the colonial exploitation of India by Britain, in the course of which vast tracts of forests were plundered to provide timber for railways, ships and furniture, to the continuing destruction of forests throughout the developing world to provide arable land to grow cash crops for export to pay off debts to the international banks.

In North America and Europe environmental movements have successfully challenged some of their societies' excesses. In the USA these date from the publication of Rachel Carson's *Silent Spring* in 1962, though there were antecedents in the nineteenth century when Ralph W. Emerson (1803–82) and others argued for the creation of national parks. It is interesting to note, in passing, that these New England Transcendentalists, as they were known, used religious arguments in support of their proposals.

During the 1960s and early 1970s environmental groups proliferated and brought about a number of improvements. Clean air and water acts were introduced and environmental regulatory agencies were set up. By the late 1970s and 1980s large corporations in North America, Europe and Australia had organized themselves to block further environmental legislation, though in the 1980s public concern began to rise again as new scientific discoveries were made about ozone depletion and global warming. The failure of the post-Kyoto discussions of global warming at The Hague in 2000 represents a major setback.

The last decade has witnessed increased confrontation between environmentalists and the large corporations, which have resorted to the law to serve intimidatory lawsuits on individual protesters. The ideology of limitless consumption has become their dominant creed, as the following claim by a prominent sales analyst in the USA makes clear:

> Our enormously productive economy . . . demands that we make consumption our way of life, that we convert the buying and use of goods into rituals, that we seek spiritual satisfaction, our ego satisfaction, in consumption. We need things consumed, burned up, worn out, replaced and discarded at an ever increasing rate.[1]

Although the environmental organizations have heightened public awareness of environmental issues, there are signs that they have lost their sense of direction. Instead of attacking irresponsible multinationals, or governments which deploy nuclear weapons, they destroy genetically modified crops and protest against nuclear waste from power plants. From the standpoint of the poorer Asian societies, the destruction of any food crops, whether genetically modified or not, is irresponsible, and from the standpoint of India, nuclear power (and therefore the safe disposal of nuclear waste) is an integral part of national energy planning.

The Asian viewpoint

In the so-called Third World as a whole, between 1960 and 1992, life expectancy at birth rose from 46 to 63, infant mortality declined by more than half, and real per capita income almost trebled.[2] These benefits were not shared equally, but the fact that they occurred was completely at variance with the pessimistic predictions of economists in the 1960s.

Many western analysts believe that the world's economic and environmental problems could be substantially reduced if only the developing world would follow China's 'one child policy'. This view has been challenged by Amartya Sen. Comparing China with India, Sen points out that whereas their population growth rates are 1.4 per cent and 2.1 per cent respectively, the growth rates of their per capita Gross Domestic Product (GDP) would change very little if these percentages were exchanged. In other words, if China had a population growth rate of 2.1 per cent (which is India's), its per capita GDP would only decline from 7.7 per cent per year to 7.0 per cent. (This assumes no change in the growth rate of total GDP.) Similarly, if India's population growth rate was 1.4 per cent (which is China's), the growth rate of per capita GDP would only increase from 3.1 per cent to 3.8 per cent.[3]

In spite of such claims – in this case by a Nobel Prize winning economist – many people continue to believe that the 'population crisis' is largely responsible for many of our global economic and environmental problems. Such people often choose to disregard a different kind of 'population crisis', i.e. that every child born in an industrial nation consumes eight times as much of the earth's natural resources as a child born in a developing country.[4] Furthermore, we might do well to ponder the fact that if the entire population of the world were transported to the USA and spread evenly across the states, the overall population density there would be no greater than it is now in the Netherlands!

The issue of population growth is capably handled by the United Nations report on environment and development that paved the way for the 1992 Earth Summit. It was published in 1987 under the title *Our Common Future*. The report acknowledges that in order to achieve sustainable development the world's levels of population increase must decline, but links this to the promotion of women's rights:

> A population policy should set out and pursue broad national demographic goals in relation to other socio-economic objectives. Social and cultural factors dominate all others in affecting fertility. The most important of these is the roles women play in the family, the economy, and the society at large. Fertility rates fall as women's employment opportunities outside the home and farm, their access to education, and their age of marriage all rise. Hence policies meant to lower fertility rates not only must include economic incentives and disincentives, but must aim to improve the position of women in society. Such policies should essentially promote women's rights.[5]

The report also acknowledges the potential role of the world's major religions in addressing social and environmental issues:

> Sustainable development requires changes in values and attitudes towards environment and development – indeed, towards society and work at home, on farms, and in factories. The world's religions could help provide direction and motivation in forming new values that would stress individual and joint responsibility towards the environment and towards nurturing harmony between humanity and environment.[6]

We shall return to this prospect presently.

India's entry into international environmental politics has been documented by O. P. Dwivedi, M. G. Rajan and George A. James.[7] It began with

Mrs Indira Gandhi's speech at the United Nations Conference on the Human Environment at Stockholm in 1972. She was the only head of state from outside Sweden to attend. In her speech she acknowledged the wanton destruction of forests and wildlife, but also drew attention to the problem of meeting the needs of the poor:

> When they themselves feel deprived, how can we urge the preservation of animals? How can we speak to those who live in villages and in slums about keeping the oceans, the rivers and the air clean when their own lives are contaminated at the source? The environment cannot be improved in conditions of poverty.[8]

During the next two decades India took part in the major international environmental debates which led to the 1992 UN Earth Summit in Rio de Janeiro with varying degrees of enthusiasm. But there were two things which set India apart from many other participants. The first, shared by some developing countries, was the recognition that the UN environment and development agenda had from the start been set by western nations which were much more interested in ozone layer depletion and global warming than in deforestation, water resources and natural resource depletion, which are crucial for India. The second, shared to some extent by Thailand, was that there are cultural resources which can be brought to bear on these problems and correspondingly appropriate ways of addressing them.

During the nineteenth century there had been sporadic opposition to colonial forest policies in India, and even a century earlier the Bishnois had protected trees from being felled by their rulers by hugging them. But it was in the 1970s and 1980s that the Chipko and Appiko movements gained momentum.[9] The 1984 Bhopal disaster demonstrated the dangers of forcing the pace of agricultural production without exercising sufficient care over the manufacture of potentially dangerous chemical fertilizers and pesticides.

Ecology

Ecology is the study of relationships among organisms and between them and their environment. Our primary concern is with the relationships between people, other organisms such as plants and animals, and the natural environment.

As an academic discipline, ecology is located among the biological sciences. Whereas biochemistry, cell biology, histology and the anatomical sciences relate to the foundational levels of biological organization (i.e. structural: organic, cellular and molecular; and functional), ecology deals with populations, communities and ecosystems. It is therefore a bridge between the biological and the behavioural sciences. The term comes from two Greek words, *oikos*, meaning 'home', and *logos*, meaning 'understanding'. Ernst

Haeckel, a nineteenth-century German who invented the term, described ecology as 'the domestic side of organic life'.[10]

The scientific study of the environment can be undertaken by dividing it into four segments: the atmosphere, hydrosphere (e.g. oceans), lithosphere (e.g. rocks, soil) and the biosphere (parts of the atmosphere, hydrosphere or lithosphere where life exists). Each segment can be studied from the perspective of the conventional sciences, such as physics, chemistry, botany, zoology and ecology. The University of Delhi recently introduced environmental chemistry into its honours chemistry B.Sc. degree; it took three lecturers (including the author) to cover the course!

It is our contention that whereas in the West (i.e. the 'developed' industrial nations) environment and development are often regarded as unrelated, in the developing world environmental issues are experienced as integral to their social context. Thus, for example, deforestation in south Asia is not just a matter of tree loss leading to the destruction of a range of botanical species and a reduction in the extent to which carbon dioxide can be removed from the atmosphere. It is the harbinger of a whole range of ecological issues: loss of food supplies, drinking water and other essential commodities to people and animals living in a symbiotic relationship with the forests, an increase in malaria, and additional hardships for women. No wonder the developing nations at the 1992 Earth Summit reacted with shock and anger to the bland arguments of the industrial nations that they should grow more trees primarily to mop up the excess carbon dioxide cause by the West's profligate lifestyles!

It is a moot point whether or not the West should itself isolate 'environment' and 'development' issues from one another. Clearly the UN believes that it should not. In developing countries, however, they cannot and must not be separated, and for that reason we shall consistently prefer the relational term 'ecological' to 'environmental'. Albert Einstein once defined the environment as 'everything that's not me', thereby indicating the crux of our problem, because in reality we are an integral part of the natural world. However, our technology and our cultural presuppositions – in part religious – have alienated us from our true context.

There are a number of recent publications which analyse environment and development issues from an ecological perspective. *Ecology*, by J. L. Chapman and M. J. Reiss, *Ecology and Environment* by P. D. Sharma, and the *Dictionary of Ecology and Environment* by P. H. Collin, are excellent introductory reading.[11] *Sustaining Earth*, edited by D. J. R. Angell, J. D. Comer and M. L. Wilkinson, contains useful scientific information relating to the Earth Summit, and an informative article about tropical forests by Ghillean T. Prance, the former director of Kew Gardens.[12]

Important points of view on a range of ecological issues are given by Arne Naess, Sulak Sivaraksa and Martin Palmer in *Ethics of Environment and Development*, edited by J. R. Engel and J. G. Engel, and much the same ground is covered in *Ethics, Religion and Biodiversity*, edited by Lawrence S.

Hamilton.[13] There are some notable contributions by Frédérique Apfell-Marglin and others in *Global Ecology: A New Arena of Political Conflict*, edited by Wolfgang Sachs.[14] More recent publications include *Global Ethics and Environment*, edited by Nicholas Low, and *Politics and the Environment*, edited by James Connelly and Graham Smith.[15] Asian regional publications will be discussed later.

Religion

Comparing environmental movements in the USA and India, Christopher K. Chapple observes:

> Whereas in the American context, the early rallying cry for environmental action came from scientists and social activists with theologians only taking interest in this issue of late, in India, from the outset, there has been an appeal to traditional religious sensibilities in support of environmental issues.[16]

This is also the case in Sri Lanka and parts of southeast Asia, notably Thailand.

In view of the important role played by religion, we must consider what we mean by it and the various levels at which it operates.

Religion is generally understood to mean the acknowledgement of a world-transcending reality (or realities), personal or not, in which human fulfilment may be achieved. The orthodox Judaeo-Christian and Muslim belief in God (Yahweh or Allah) is an example of a world-transcending personal reality; the Buddhist notion of *nirvāṇa* (Sanskrit) or *nibbāna* (Pali) is non-personal. It is therefore not correct to view religion as necessarily theistic in a personalist sense, because much Buddhism, large parts of the Hindu tradition and the mystical traditions of Islam, Judaism and Christianity cannot be classified so easily.

Religions are multi-dimensional, manifesting themselves in a variety of areas of life. Ninian Smart categorizes these as practice and ritual, experience and emotion, narrative and myth, doctrine and philosophy, ethics and prescription, societal and material factors.[17] Whereas western commentators tend to stress the doctrinal and ethical components of religion, in Asia the emphasis is usually much more on experience, ritual and prescriptive social behaviour.

In the following chapters we shall consider the Hindu and Buddhist traditions as they occur in south and southeast Asia. Hindus constitute 82.5 per cent of India's population, which now borders on a billion. Twelve per cent are Muslims, 2 per cent Christians (Roman Catholics, Syrians and Protestants) and 2 per cent are Sikhs. Buddhists constitute less than 1 per cent and the Jains are less than 0.5 per cent. Ladakh, which is part of Kashmir, is predominantly Buddhist, as is Bhutan, which is a separate country. (We shall consider these two Himalayan kingdoms in Chapter 5.) Ninety-five per cent of Thailand's population of 53 million are Theravāda Buddhists; the remainder are mostly Muslims in the south.

'Hinduism' is a family of culturally similar traditions, a unity-in-diversity. The culture is geographically based on the sub-Punjab area of Sind, where a river described in the Vedas as the Sindhu (Indus) flowed.[18] From the earliest times rivers possessed a transcendent power. 'Hinduism' is polycentric and cannot be reduced to any simple formula or essence; we shall therefore avoid the term in the following chapters, except in quotations, preferring 'Hindu tradition'. 'Hindu' is both a geographical and a cultural term.

Buddhism is less diverse and can conveniently be divided into major streams such as the Mahāyāna and the Theravāda, though even here we must be careful not to make the distinctions too neat. The Buddhism of the Himalayan kingdoms may be broadly described as Mahāyānist; that of the southern and southeastern regions is mainly Theravādin.

We shall have occasion to use 'culture' almost synonymously with 'religion'. A comprehensive definition of the term has been given by M. M. Thomas, a distinguished historian and former governor of Nagaland:

> Culture . . . [denotes] the structures of meaning and sacredness within which a people relate themselves to nature, organize themselves to fulfil the physical urges of sex and hunger and build up community relations on codes of justice and solidarity . . . Culture is an embodiment of values in forms capable of influencing imagination and affections, and therefore of motivating behaviour, a dimension between religious truth and social institutions. It shapes the distinctive character, the selfhood, of a people.[19]

Although many regional and religious or ethnic communities in India possess their own particular culture, there is an emerging, if fragile, sense of an inclusive national culture. In Thailand, 'Thai' and 'Buddhist' are virtually synonymous, except along the southern border with Malaysia, where there is a significant Muslim presence.

The Hindu tradition and Buddhism are considered together because they share a common history and for reasons to do with our methodology (see next section). Although there are not many Buddhists in contemporary India, Buddhism exercises a strong influence, as the following extract from Mrs Indira Gandhi's 1972 speech to the UN in Stockholm makes clear. In it she refers to the close affinity which she, as a Hindu, had always shared with the world of nature, and to the Buddhist values which had led the Indian emperor Ashoka to protect animal life and conserve forests:

> I had the good fortune of growing up with a sense of kinship with nature in all its manifestations. Birds, plants, stones were companions and, sleeping under the star-strewn sky, I became familiar with the names and

movements of the constellations... One cannot be truly human and civilized unless one looks upon not only all fellow-men but all creation with the eyes of a friend. Throughout India, edicts carved on rocks and iron pillars are reminders that twenty-two centuries ago the Emperor Ashoka defined the king's duty as not merely to protect citizens and punish wrong-doers but also to preserve animal life and forest trees. Ashoka was the first and perhaps the only monarch until very recently to forbid the killing of a large number of species of animals for sport or food.[20]

There have been a number of important recent publications dealing with the relationship between ecology and the Hindu and Buddhist traditions. J. Baird Callicott's *Earth's Insights: A Survey of Ecological Ethics from the Mediterranean Basin to the Australian Outback* explores the connections between moral and ethical values through sacred texts, drama and iconography, in an attempt to create a universally acceptable environmental ethic.[21] David Kinsley proposes a model for understanding pilgrimages and other Hindu customs as a form of ecological spirituality in *Ecology and Religion: Ecological Spirituality in Cross-Cultural Perspective*.[22]

Purifying the Earthly Body of God, edited by Lance E. Nelson, is a symposium of articles about the Hindu tradition and ecology by distinguished scholars from North America. In one of these, Christopher K. Chapple concludes a survey of Indian scriptures which have a continuing ecological relevance (including those of Yoga, Jainism and the tribal groups) with the assertion that 'through these emerging indigenous resources and resourcefulness, India will devise its own methods and rationalities in defense of the ecosystem'.[23] *Buddhism and Ecology* is part of a series of publications on world religions and ecology sponsored by the Harvard Center for the Study of World Religions.[24] Articles range from assessments of the ecological potential of Japanese Buddhism to studies of Buddhism in the USA.

Roger S. Gottlieb's *This Sacred Earth* contains selections from a number of traditional religions, and is therefore a useful source book. He maintains that in order to relate these religions to environmental problems, they need to be reinterpreted, extended, synthesized with borrowed elements, or used as the basis for creative innovation.[25]

There is a bibliography and a glossary of technical terms at the end of the book. The introductory paragraph of the glossary explains the rationale for italicizing words and phrases in the text.

Methodology

The broad scope of this study of ecology and the Hindu and Buddhist traditions of south and southeast Asia is to some extent a consequence of our methodology. It is also intended to provide a backcloth for more focused and detailed investigations into particular issues.

We shall consider the Hindu and Buddhist traditions from the point of view of both their histories, which are to some extent shared, and their contemporary expression, considering in particular what Stanley Tambiah describes as the 'continuities and transformations' between the past and the present. From these resonances between history and anthropology, between the past and the present, we may draw tentative conclusions about the role of religion in raising social and environmental awareness.

During the 1950s and 1960s social scientists working in India coined the notions of the 'little tradition' of village India and the 'great tradition' of the classical Hindu texts, and wondered why they were so different. Marriott, Redfield and the Chicago School maintained that these differences were actual, whereas Dumont and Pocock claimed that the traditions are one and the same and that their unity is based on a reiterated relationship which can be traced in different areas of Indian life.[26] Thus, they argued, while it may sometimes be legitimate to distinguish between popular Hinduism and the traditional higher Sanskritic civilization, it should be remembered that much of what properly belongs to the literary texts has been lost or omitted.

Without proceeding further with this particular debate, it will be clear that Dumont and Pocock's popular and Sanskritic levels of Hinduism can readily be paralleled with similar notions relating to Thai Buddhism. In *Buddhism and the Spirit Cults in North-East Thailand*, Tambiah develops these ideas further and applies them to his own anthropological studies carried out in the early 1960s.[27] Tambiah's work has been hailed by Jane Bunnag as 'the first really solid anthropological study of any aspect of Thai society' and it is not without good reason that he was appointed to Mircea Eliade's chair at Chicago University prior to his current position as head of anthropology at Harvard.[28]

Tambiah criticizes the notion of two levels in religion – the higher literary and the lower popular – on the grounds that it is 'in some respects static and profoundly *a-historical*'. In fact, most religions are constantly changing both their beliefs and structures, and the classical texts, whether in Sanskrit or in Pali, 'range over long periods of time and show shifts in principles and ideas'.[29] He therefore redesignates the higher and lower levels of religion as follows:

> In the study of religion in societies like Thailand, I would make a distinction between *historical* religion and *contemporary* religion, without treating them as exclusive levels. Historical Buddhism would comprise not only the range of religious texts written in the past, but also the changes in the institutional form of Buddhism over the ages. Contemporary religion would simply mean the religion as it is practised today and should include those texts written in the past that persist today, and are integral parts of the ongoing religion. Thus, if the question of the relation between historical and contemporary religion interests us, we should look for two kinds of links, namely, *continuities* and *transformations*.[30]

10 *Religion and Ecology in India and Southeast Asia*

I shall give examples from the Hindu and Buddhist traditions of continuities and transformations of ecological significance between the past and the present in the following chapters. In order to clarify our methodology, however, I shall offer a recent illustrative example from the Judaeo-Christian tradition.

Christians who wish to demonstrate that their religion is eco-friendly may do so by citing naturalistic passages from their scriptures; for example: 'Think of the ravens: they neither sow nor reap ... yet God feeds them' (Luke 12.24, *New English Bible*). Clearly, by a process of simple extrapolation from the past to the present, birds should be valued by Christians. This is a straightforward continuity between the past, as reflected in canonical scripture, and the present.

An example of a transformation can be seen by the manner in which Christians have adapted the notion of 'jubilee' to endorse the cancellation by western banking institutions of the debts of poor countries. This is much more than a simple extrapolation of the Jewish practice of restoring land and releasing bonded labour every fifty years. The environmental implications of debt cancellation were first worked out by a section of the World Council of Churches in the mid-1980s, and were subsequently taken up by a number of western churches and agencies.[31] The justification for cancelling the debts is valid with or without the appeal to jubilee, but the symbolism has added a powerful religious dimension to the campaign.

Main arguments

Recent surveys indicate that the vast majority of young Indians between the ages of sixteen and thirty believe in God and/or consider themselves to be religious.[32] In Thailand the majority of young men are ordained at least temporarily as monks, and women are increasingly able to enter religious vocations as lay nuns. For all their secular veneer, both Indian and Thai societies are deeply sensitive to some or other aspect of religion and local culture.

This study considers both the historical and the contemporary dimensions of the Hindu and Buddhist (mainly Theravāda) traditions from the point of view of ecology. This means that we must enlarge their conventional frameworks to bring out what Bridget and Raymond Allchin describe as 'the ecological relationship between a human community or group and its environmental context'.[33]

It is important to stress that we are rediscovering an existing framework and not imposing a modern one on the historical material. Thus in Chapter 2 we consider an account of periods of early history from the point of view of various modes of resource use: hunter-gatherer, nomadic pastoralist and settled agriculturalist (not necessarily in chronological order). The industrial extractive mode, with all its devastating environmental consequences, is described later.

Chapter 2 also contains examples of episodes from the Vedic literature which lend themselves to an ecological interpretation. For example, when the god of fire, Agni, is described as 'eating up' the forests to the east, we conclude that the northern invaders were burning forests in order to create settlements as they moved from what is now the Punjab into the central Gangetic plains. We consider some of the pre-Vedic strands of the Hindu tradition such as the Indus civilization, and cite some of the more naturalistic literary passages, such as the Bhūmi Sūkta (Hymn to the Earth).

Not all the historical literature supports the romantic view of some scholars that the Hindu tradition unreservedly endorses ecological harmony between people and nature.[34] Thus the razing of the Khāṇḍava forest by Krishna and Arjuna (Chapter 2) may not in itself be ecologically irresponsible, but the wanton manner in which they fling the escaping wildlife back into the flames, laughing and joking as they do so, is embarrassing to those who wish to present the Hindu tradition as a model of ecological propriety. In spite of such blemishes, however, there is a convergence among the various components which have given rise to the tradition as a whole that considerable affinity exists between humans and nature – so much so that even trees are 'pervaded by *ātman*' (i.e. self). We conclude Chapter 2 with a summary of the main tenets of the Hindu tradition incorporating its ecological dimension by Dr Karan Singh, a distinguished contemporary scholar.

Chapter 3 describes some of the consequences of the industrial era. During this period a range of scientific, technological and industrial influences entered India from the West and set in motion irreversible changes. Forest ecosystems were irreparably damaged as trees became commodities, and educated India underwent a cultural transformation which resulted in a broad range of social and religious reforms and also led to a re-evaluation of traditional philosophy and science. We consider Vivekananda's view that communality exists between all life forms, and Gandhi's rejection of anthropocentrism. The chapter concludes with an account of Indian scientists who directed their scientific research according to their Hindu beliefs: thus Jagadish Chandra Bose's pioneering work on the possibility of pain in plants represents a remarkable transformation of tradition (in Tambiah's sense), and suggests the prospect of more ecologically and culturally sensitive forms of science – a point taken up again in the final chapter.[35]

Chapter 4 explains how the economic policies of both colonialism and post-Independence India eventually led to resistance to commercial forestry in the Himalayas. These struggles were not 'environmental' in the sense of, say, attempts in Europe to prevent the destruction of a forest in order to expand an airport, but responses to a total threat to communities living in a symbiotic relationship with their natural surroundings. Events leading up to the Chipko movement are described with particular reference to the use of traditional symbolic activities such as the *padayātrā*, readings from the Bhagavadgītā, etc. The emergence of Gandhian, Marxist and 'appropriate technology' wings within Chipko is noted and accounts are given of the work of Sunderlal Bahuguna and Anna Hazare and the influence of Vivekananda, Gandhi and Jagadish Chandra Bose on them.

Buddhism appeared in India well after the three major stages of resource use which have been described. Unlike the Hindu tradition, with its gods and their intermediaries, it began as a view of human existence with implications for human society, and as a philosophy with no need of theistic beliefs or sanctions, yet tolerant of them. Chapter 5 considers the rise of Buddhism from the birth of the Buddha in a sacred grove of *sal* trees until and beyond its development into the Theravāda and Mahāyāna traditions. The Theravāda built upon early scriptural parallels between the Buddha and the king, which received tangible expression in Ashoka's understanding of himself as *cakravartin*, turner of the wheel of *dharma*, and as the one through whom all *dharma* flows. Responsibility for all life, including wildlife and forests, therefore becomes the duty of the king. These notions were exported to Sri Lanka and southeast Asia, where they provided an ecumenical framework from within which social and environmental issues could be addressed. The Mahāyāna took root in northern India and in the Himalayan regions, where it acquired a strong naturalistic emphasis because of such notions as the *bodhisattva*, who refuses to enter *nirvāṇa* until all sentient beings can do the same, and Tibetan naturalism. Ladakh and Bhutan are considered as regions where this stream of Buddhist tradition has been transformed into a potent vehicle for ecological improvement.

Chapter 6 introduces Thailand as an example of an Asian society which, because it has not been colonized, has been able to maintain its monastic and cultural traditions in unbroken continuity from the past. Although not formally colonized, however, Thailand has experienced many of the problems of resource depletion shared by other southeast Asian societies, particularly deforestation and the destruction and pollution of waterways. Since the 1970s an increasingly influential 'green lobby' has drawn attention to the severity of these problems and the inappropriateness of government 'development' in terms of large-scale dams and eucalyptus plantations. Protesters have included farmers, environmental non-governmental organizations (NGOs) and Buddhist monks who have resisted the clear felling of forests by 'ordaining' trees and creating sacred groves with saffron cloth. Much of this chapter is based on fieldwork I have conducted.

Chapter 7 considers India's post-Independence model of centralized planning from the point of view of equitable development and the state of the environment. India's participation in international environmental politics since the 1970s is reviewed, and consideration is given to various proposals for ameliorating social and environmental problems. These include the policies of the Hindu Right, currently represented by a BJP-led coalition government, and the recommendations of Gadgil and Guha[36] compared with those of Drèze and Sen.[37] The former of these pairs advocates an environment-led approach based essentially on a combination of the philosophies which emerged during the Chipko movement (e.g. Gandhian, Marxist); the latter urges participatory development at local and district levels (backed by strong central support), concentrating on basic education, healthcare and a reduction of the gender imbalance.

Chapter 8 reviews positive initiatives ranging from those of Sunderlal Bahuguna and Medha Patkar at the national level and the collective representations of the 100 Indian NGOs which took part in the Earth Summit to those of a growing number of individuals in different parts of India who are using culturally appropriate methods to address social and environmental problems. Chapter 9 reviews the earlier ones in terms of the 'continuities and transformations' between the past and the present which appear to offer the greatest potential for ecological improvement.

Religion at Rio

We conclude our introductory remarks by considering the remarkable degree to which Indian NGOs and both the Indian and Thai governments participated in the Earth Summit held in Rio de Janeiro in 1992, and the extent to which they used religious arguments to press their cases.

We have already noted the fact that environmental priorities as seen from the standpoint of developing societies are very different from those of the West. This point has been emphasized in relation to the Earth Summit by Walter Fernandez, director of the Indian Social Institute in Delhi:

> At the Rio Summit in 1992 . . . groups . . . ensured that mainly issues like global warming that affected the rich nations were dealt with and that important aspects like desertification that affected the poor were ignored.
> . . . The same groups would also speak of the traditional practices as destructive. For example, they attribute deforestation mainly to shifting cultivation and overpopulation.[38]

Be this as it may, the Earth Summit did address a number of important issues and provided an opportunity for an enormous number of participants to exchange views. The Summit itself was divided into two simultaneous events. The main international proceedings, attended by heads of state, were held some distance outside Rio de Janeiro. There was also a forum for NGOs in downtown Rio. However, many NGOs were also accredited to the main conference; these had submitted a statement of their aims, membership and other data in advance for approval by the UN.

Cultural and religious contributions to the Summit were possible either at state level by, for example, the Vatican, or via NGOs. We shall consider here and again in more detail in Chapter 8 only those NGOs which had been vetted in advance and accredited by the UN.

Considering first the state-level contributions to the main conference, the Vatican submission raised important questions about an issue that was otherwise largely ignored, namely the levels of consumption and waste of the industrial nations:

> Complementing respect for the human person and human life is the

responsibility to respect all creation. God is creator and planner of the entire universe. The universe and life in all its forms are a testimony to God's creative power . . . The scandalous patterns of consumption and waste of all kinds of resources by a few must be corrected in order to ensure justice and sustainable development to all everywhere in the world.[39]

The Thai Government presented its arguments in characteristically Buddhist terms. According to Princess Chulabhorn, who led their delegation:

There is a need to agree to a fundamental set of principles that will provide a critical framework for our actions as individuals as well as collectively as nations. These principles must revisit the need for equity, justice, peace, compassion, respect and care between human beings and all the manifestations of nature, the priceless heritage for all that share this unique, beautiful and bountiful world.[40]

It was the Indian Government's official report, *Traditions, Concerns and Efforts in India*, however, which had most to say about religion.[41] It included quotations from the Upanishads and references to sacred groves, described as 'a unique tradition . . . for preserving pockets of biodiversity'. A pillar edict by Ashoka was quoted: 'Forests are not to be burnt, either uselessly or for killing [animals]'. There were references to the Chipko and Appiko environmental movements, and the back cover of the report quoted Mahatma Gandhi: 'The earth has enough for everyone's need but not for anyone's greed.'

Of the 1,400 NGOs officially accredited to the United Nations conference, exactly 100 came from India. This is a remarkable level of representation compared with, say, less than a dozen each from the smaller countries adjacent to India. A full list of all these organizations is given in Appendix B, together with their locations and a summary of their aims. It is clear from this list that very few organizations, other than large ones such as the Centre for Energy and Environment and the Tata Energy Research Institute, located in Delhi, paid any attention to ozone depletion and global warming. Most were much more concerned about deforestation, waterway and other types of pollution, healthcare, women's education, resource depletion and the basic needs of predominantly rural communities. Though classified by the UN as NGOs, most were essentially concerned citizens' or people's groups. Several of them described their aims in cultural or religious (largely Gandhian) terms. A map of India's states is given at the end of this chapter to assist in the location of the various NGOs.

It seems pertinent to ask why there was such a high level of participation from Indian representatives at the Earth Summit, and why several of them and two of the south and southeast Asian official delegations should have expressed their views in such unequivocally religious terms. In the following chapters we shall consider these questions and the potential of the Hindu and Buddhist traditions for addressing major social and environmental problems.

Figure 1.1 India's states. The three new states which came into being in October 2000 are small and have been omitted.

2 Ecology and Hindu tradition

Supporters of the view that the Hindu tradition inculcates respect for the natural world might be surprised to know of a passage in one of the great epics which advocates the wanton destruction of plant and animal life. The episode occurs in the Mahābhārata, which was composed over a long period between 800 BCE (Before the Common Era) and 400 CE (Common Era) and is about the trials and tribulations of the Bhārata clan.

The story takes place in the kingdom of Kurukṣetra which is situated between the Ganges and Yamuna rivers, and much of it concerns the struggles for supremacy between the Kauravas and the Pāṇḍavas.[1] The latter live in the Khāṇḍava forest, which contains their capital city of Indraprastha (now Delhi). Krishna and Arjuna are in the forest one day when a poor brahmin asks for alms. They grant his request, whereupon he is transformed into Agni, the god of fire, who insists that his hunger will only be satisfied if the entire forest is consumed by him.

As the forest begins to burn the forest creatures, including *nāgas* (snakes, usually cobras) flee the flames. The two warriors race round and round the blazing forest in a chariot, catching the escaping wildlife and hurling it back into the flames. As they do so they laugh gleefully and joke with one another.

The episode has been interpreted as a great Vedic sacrifice intended to placate Agni. More realistically it probably reflects the clearance of forests to provide land for settled cultivators. If this is the case, then the fleeing *nāgas* are the forest hunters or tribals, who worshipped snakes. The fact remains, however, that Krishna and Arjuna are among the most important figures in Hindu literature (e.g. the Bhagavadgītā) and their wanton cruelty towards animals in this episode remains something of an embarrassment to those who would like to present the Hindu tradition as a streamlined corpus of ecological propriety.

In this chapter I argue that in spite of such problematic passages in the scriptures, there is a convergence among the various strands that make up the tradition as a whole that human life and nature are part of a single continuum. I shall do this by considering the attitudes of the earliest human communities to nature, characterizing them according to their different modes of resource use. I shall explore the possibility that the literary tradition reflects, in part at least, some of the changes which occurred in relation to these (e.g. settled cultivators clearing forests). I shall offer a brief account of the most important elements of the Hindu tradition, stressing those aspects which seem to have an ecological significance.

At the conclusion of this chapter, in order to put some flesh on an otherwise very sketchy outline, I shall summarize the views of an ecologically minded modern Hindu. These views represent continuity with the past – in Tambiah's sense – and reflect an ecological dimension that was originally part of the tradition. I am not imposing contemporary environmental concerns on the distant past.

Modes and habitats

The importance of taking into account the ecological relationship between human communities and their environmental contexts has been described by Bridget and Raymond Allchin:

> During the last two decades South Asian prehistory has moved into a new phase. Previously the majority of prehistorians working in the subcontinent were concerned with artefact typology and technology. . . . Recently, owing both to new approaches beginning to be adopted by prehistorians working in the subcontinent and to advances in related fields, particularly geomorphology, palaeontology, palaeobotany and palaeoclimatology, throughout the world, there has been an increasing swing towards considering past cultures in their totality. This means finding out as much as possible about the ecological relationship between a human community or group and its environmental context.[2]

In this and the following sections we shall consider early Indian history from the point of view of the various modes of resource use. These are: hunter-gatherer, nomadic pastoralist and traditional settled agriculturalist. We shall consider the industrial urban mode and modern 'developed' agriculture later.

The first mode, which is the longest in human history, is characterized by the hunting of wild animals and the gathering of vegetable matter for subsistence. Energy is supplied by human muscle power and fuelwood. Women gather plant foods and small animals; men hunt and are collectively responsible for decisions within social kin groups typically composed of between ten and 100 members.

Many of the practices of these hunter-gatherers continue to be adopted by tribal groups in various parts of India. These include the conservation of resources according to seasons (e.g. certain animals may not be hunted in many Indian villages between July and October) and restrictions on the cutting of particular species of tree such as *sal* (*Shorea robusta*). These trees are regarded as sacred, as are clusters of trees surrounding burial grounds. Clusters of *sal* trees on the outskirts of villages were known as sacred groves.

Hunter-gatherers experience nature as capricious and beyond human control, but they regard themselves as part of a community of animals and plants, mountains and rivers. Their religion is centred around spirits, gods and demons which inhabit trees, rocks, birds, etc. and must be worshipped and occasionally placated.

Tribal groups in India, as elsewhere, usually subscribe to myths centred on the creation of the world or the origin of the tribe. Some such myths strongly advocate the protection of natural objects, such as trees, on account of their usefulness. Thus the Didayis of Orissa believe that when the world was destroyed by a flood, the supreme deity made a new world without trees. Consequently people could not cook, build houses or find shelter. When God saw their suffering he added trees.

The ecological and religious importance of trees from the earliest stages of human existence cannot be overestimated. According to Vandana Shiva:

> The protection and propagation of forests as a deeply ingrained civilizational characteristic in the South Asian region is evident from the existence of sacred groves in river catchments ... and from village woodlots. The practices were of critical value both ecologically and economically. Ecologically, indigenous and naturalized vegetation has provided essential life support by stabilizing the soil and water systems. Economically, trees have been a source of small timber, fodder, fuel, fibre, medicines, oils, dyes, etc. Indigenous medicines use more than 2,000 species of plants, both wild and cultivated. The centrality of trees to survival and economic well-being created the need for their conservation which was achieved through the concept of sacredness.[3]

Nomads and settlers

The hunting and gathering mode of resource use came to an end in many places as plants and animals became domesticated. However, animal husbandry must be based on the movement of herds according to the seasonal availability of land for grazing. This meant that the successors of the earliest hunters and gatherers were nomads who relied on animals for transport and their products for food. Their mobility gave them little opportunity for attachment to sacred places; if resources became scarce they mounted their animals and moved elsewhere. Eventually many of them began to settle.

Settled cultivation involves a form of energy use in which human muscles are supplemented with fuelwood and animal and water power. Land must be cleared so that certain species of plant – wheat, rice and maize, for example – can be intensively cultivated. Small kin groups are more viable in settled cultivation than larger ones, with the result that families tend to become the basic units of agricultural societies. However, families need to band together for mutual support in villages, which become the larger functional units. Men plough, women tend plants, collect fuel and water, clean and grind grain and cook. Forests, grazing ground and water are commonly owned by each village. Where villages have become part of larger territories there may be considerable division of labour and the emergence of specialized groups, such as priests, who are responsible for natural and cultural knowledge, and warriors, who protect the interests of the village or group of villages from outsiders.

Agricultural societies occupy something of an intermediate position between hunters and nomads in their relationship with nature. On the one hand, unlike the hunters, they have established a firm measure of control over their surroundings; on the other, unlike the nomads, they cannot merely mount their horses and go elsewhere when nature becomes capricious. They therefore appear to share with the hunters and gatherers a sense of being part of a community of life, and with the nomads a strong sense of their ability to control resources. According to Madhav Gadgil and Ramachandra Guha:

> The restrained use of natural resources could thereby be expected to form one part of the ideology of agricultural societies, especially when they are in a state of near equilibrium with their resource base. On the other hand, agricultural societies in the process of encountering an expanding resource base – either through new technologies or, especially, while colonizing lands earlier held by gatherers – are much more likely to view man as separate from nature and with a right to exploit resources as he wishes.[4]

We are now in a better position to interpret some resource conflicts between representatives of different modes of resource use. We considered one at the beginning of this chapter where Arjuna and Krishna, representing settled cultivators, destroyed a forest to provide more land. Another relates to the early history of the Mundas of Jharkand; it is known as the Asur legend.[5]

The Asur spirits in this legend are the tutelary spirits of the Mundas and other related tribes in the Jharkand area of Bihar. There is one supreme spirit, known as Singbonga, who created and sustains all things and various lesser spirits, many of which are ancestral. Each village is protected by Desauli Bonga, the spirit of the sacred grove.

According to the Asur legend, certain Asurs who were iron smelters burned their fires so persistently that vegetation was scorched, water supplies dried up and the air was polluted. This displeased Singbonga, who burned all the Asurs to death in a furnace. The women then complained that without their menfolk they could not survive. So Singbonga took the charred bones of the men from the furnace and scattered them all over the earth. Falling on mountains, rocks, deep waters and wooded places beside springs, the bones became *bongas*, or guardian spirits of those places. Thus

Singbonga resolved a conflict between a settled tribe of cultivators who had resorted to environmentally unsound practices and the original inhabitants by restoring a truly symbiotic relationship between his people and nature.

We have considered the earliest stages of history in relation to three modes of resource use (and it is important to note that these may have to some extent coexisted). The fourth mode of energy use is industrial, and follows an extractive route whereby natural resources are mined (e.g. coal) or harnessed (e.g. hydro-electric power) for human consumption. This mode did not occur until comparatively recently and will be considered later.

Towards the end of the Pleistocene period (c. 8000 BCE) the landscape of north and northwest India was almost certainly very different from what it is today. The ability of the coastal belt of western India and the Ganges–Yamuna region to support intensive agriculture depends on monsoon rains, which owe their existence to the juxtaposition of the Himalayas, the mountain ranges of peninsular India, the north and northwestern plains and the surrounding oceans. From a geological point of view, however, the Himalayas are of comparatively recent origin and their upward movement is likely to have had a significant effect on past climates. According to the Allchins:

> The whole question of interpretation of the South Asian evidence is complicated by the rapid uplift of the Himalayas, Tibetan plateau, Pamirs and other mountain ranges . . . The rapidity, scale and recent date of this activity has only become apparent since the role of plate tectonics or 'Continental Drift' as a major factor in the shaping of these regions has been recognized. The relationship of past climates to such major tectonic activity is highly complex. At present all that can be said is that these changes must have had profound and far reaching effects not only upon the mountain regions but upon the whole subcontinent.[6]

Such climatic changes as occurred would have had their effect on major rivers, in some cases enabling them to support human communities, in others changing their course to the detriment of the same settlements. Factors such as these may have played a significant part in the rise and decline of the Indus civilization.

The Indus Valley civilization began early in the third millennium BCE. There is evidence that these settled cultivators used the plough or a similar instrument to increase their range of crops, with the result that agricultural surpluses enabled towns to flourish. New trades developed and the need to keep records led to the first attempts at literacy. The Indus city-dwellers of Harappā and Mohenjo-daro appear to have been divided into social groups which included workmen, craftsmen, traders, priests and menials. Damodar Kosambi claims that the more powerful social groups were able to induce people to part with their surplus cereal crops through a mixture of trade and religion, thus maintaining the status quo.[7]

The reconstruction of the religious beliefs of the Indus people is based on female terracotta figurines, thought to be popular versions of the Great Mother Goddess, and depictions on seals of a male deity similar to the later Śiva. Some of the seals show the male figure meditating in the position of a *yogin*, wearing a buffalo crown and surrounded by animals. He has three faces, and in some representations a plant sprouts from between his horns. Similarities between this deity and others, such as those depicted on Elamite seals, have been discussed by Gavin Flood.[8]

The Allchins summarize the main reasons for the rise and decline of the Indus civilization as follows:

> The Indus civilization arose as a social, economic and cultural phenomenon, produced by the build-up of population on the fertile Indus and Punjab plains. The resultant urban society was a delicate balance of internal relations between cities, towns and villages, and of external relations with neighbouring peasant societies and more distant urban societies. The end of the Indus urban phase probably arose from some major upsetting of this balance.[9]

The urban phase probably ended towards the close of the second millennium BCE.

The early Vedic period

The people who first spoke the Sanskrit language used the Vedas as their basis for religion and worshipped a group of male and female deities led by Indra. They called themselves *ārya*, i.e. noble, and were highly versatile warriors who used the horse, which was not initially ridden but hitched to a fast chariot. Essentially patriarchal and pastoral tribespeople, the Aryans gradually expanded their sphere of influence eastwards into the Gangetic valley. They introduced new patterns of production with the result that small tribal groups and clans were drawn into a variety of different types of social organization.

The Aryans brought with them a religion based on a sacrificial ritual (*yajña*) enshrined in collections of Sanskrit texts, some of which were in the form of hymns to deities. These underwent subsequent development until about 800 BCE, by which time they had assumed a canonical bloc of four Saṁhitās ('collections'). The first of these Saṁhitās, the Ṛk, gives the most comprehensive account of the Vedic deities: Indra (who replaced the somewhat remote Varuṇa), Agni, Soma and many others. On its own it is usually referred to as the Ṛgveda. The Sāma Saṁhitā (Sāmaveda) is much the same as the Ṛgveda.

The Yajus Saṁhitā (Yajurveda) has been transmitted in several recensions, the two main groups being known as the Black and White Yajurveda. The White Yajurveda includes the Śatapatha Brāhmaṇa, which contains revealing information about the Aryan method of land-clearing:

Mādhava, the Videgha, was at that time on the [river] Sarasvatī. Agni thence went burning along this earth towards the east; and Gotama Rahūgaṇa [the priest] and Videgha Mādhava [the king] followed after him as he was burning along. He burnt all these rivers. Now that [river] which is called Sadānīrā ['always with water'] flows from the northern [Himālaya] mountain; that one he did not burn over. That one the brahmins did not cross in former times, thinking, 'it has not been burnt over by Agni Vaiśvānara'. Nowadays, however, there are many brahmins to the east of it. At that time it [the land east of the Sadānīrā] was very uncultivated, very marshy, because it had not been tasted by Agni Vaiśvānara. Nowadays, however, it is very much cultivated, for the brahmins have caused Agni to taste it through sacrifices. Even in late summer that [river] rages along; so cold is it, not having been burnt over by Agni Vaiśvānara. Mādhava the Videgha then said [to Agni], 'Where am I to abide?' 'To the east of this [river] be thy abode,' said he. Even now this [river] forms the boundary between the Kosalas and the Videghas; for these are the Mādhavas.[10]

Thus the Aryans burnt the forests along the Himalayan foothills as they moved eastwards.

The Atharvaveda was eventually ranked equal to the other three Vedas. It contains some fine descriptions of the relationship between humans and nature:

The earth, which possesses oceans, rivers and other sources of water and which gives us land to produce foodgrains and on which human beings depend for their survival – may it grant us all our needs for eating and drinking: water, milk, cereals and fruit.[11]

The Atharvaveda also contains the Bhūmi Sūkta (Hymn to the Earth).

To each of the four Vedas was attached a Brāhmaṇa, which commented on and explained the sacrificial rites. These were followed by Āraṇyakas, forest treatises in prose, and Upanishads, which interiorized the ritual. The entire corpus of four authoritative Vedas took from approximately 1200 BCE to 500 BCE to develop.

The Ṛgveda represents the earliest stages of the Aryan presence in north India. The main concern of the Aryans was with the performance of sacrificial ritual to secure material benefits such as health, possessions, victory and immortality. It was the bridge between the empirical world and the world of the gods. As sacred fire, Agni occupied a special position with regard to the sacrifice. He was also the destroyer of forests and cities.[12]

Indra was a powerful warrior god and his crushing *vajra* ('mace', later 'thunderbolt') destroyed enemies and shattered cities. It also released the rains, however, as the monsoon clouds curled upwards like a huge dragon waiting for the first flash of lightning. According to Kosambi, it destroyed the dams constructed by the Indus people, thereby unblocking the rivers.[13]

Despite their destructive capabilities, Agni and Indra have benign characteristics. There is a type of monotheism sometimes called henotheism underlying the hymns in honour of them:

> I magnify Agni, the domestic priest, the divine minister of the sacrifice, best bestower of treasure.[14]

> Agni I consider my father, my kinsman, my brother and my friend for ever. I honour the holy light of the sun in the sky as the face of the divine.[15]

> With the waters – Soma has told me – are all medicines that heal and Agni, who blesses all. The waters contain all medicines.[16]

> O Indra! Bestow on us the best treasures: the efficient mind and inner lustre, the increase of wealth, the health of bodies, the sweetness of speech and the fullness of days.[17]

> They speak of Indra, Mitra, Varuṇa, Agni . . . The One the wise call by many names.[18]

The performance of the sacred ritual and the knowledge required to do so came increasingly under the control of professional priests, or brahmins. The original Vedic fire-priest was the *atharvan* and there were other sacrificial priests such as the *hotṛ*. These had their counterparts in Aryan societies outside India. It is interesting to note that for the correct performance of ritual during the early Vedic period it was essential for a wife to sacrifice with her husband.

Brahmins have no exact equivalent outside India; it is possible that they were the result of interaction between Aryan and Indus priests. There is some evidence for non-Aryan brahmins in the Ṛgveda and in much later *tantric* texts. The four-fold *varṇa* system, consisting of brahmin (or *brāhmaṇa*), *kṣatriya, vaiśya* and *śūdra,* has been interpolated into the Ṛgveda in the context of the account of the sacrifice of the primordial man and it is only in the Yajurveda that it is fully developed.[19]

Unlike the Ṛgveda, which mostly relates to the early period of Aryan immigration, the Yajurveda deals with regular settlements. In spite of its preoccupation with ritual, it contains some beautiful descriptions of nature:

> In spring the winds blow coolly like water, the rivers and ocean flow calmly and medicinal herbs are filled with sweet juice . . . In spring let trees give us sweet fruits, the sun physical strength and cows sweet milk.[20]

> May there be peace in the celestial regions, may there be peace in the

atmosphere, may peace reign on earth, may the waters be soothing, may the medicinal herbs be healing and may the plants be a source of balm to all.[21]

By the time of the Yajurveda the early Vedic gods had been supplemented by an all-pervasive One:

The loving sage sees that being, hidden in mystery,
Wherein the universe comes to have its home,
Therein unites and therefrom emanates the whole.
The omnipresent One is warp and woof to all created beings.[22]

The person who sees all living and inanimate creation in God and God pervading all objects does not fall a prey to doubt.[23]

The Upanishads

The Upanishads disclose the inner meaning of the sacrificial ritual. They reveal the true nature of *Brahman*, the underlying unity and reality of all diversity, of our true self (*ātman*) and of the inner relationship between them.

The Upanishads press many of their important truths through 'homology', the realization of correspondence between the inner self (or microcosm) and external reality (or macrocosm). Thus in the Bṛhad-āraṇyaka Upanishad, which forms part of the Śatapatha Brāhmaṇa, we read that 'the year is the body of the sacrificial horse'.[24] The horse's body is an aspect of its *ātman* and the year is a corresponding aspect of external reality. There is an important passage in the Chāndogya Upanishad in which a sage draws attention to a series of external realities (a tree, for example), explaining in each case that 'that which is the finest essence – this whole world has that as its soul. That is reality. That is *ātman*. That *you* are'.[25] Different interpretations of this passage gave rise to the schools associated with the names of Śaṁkara, Rāmānuja and Madhva.

In the older Upanishads, such as the Bṛhad-āraṇyaka and the Chāndogya, the early Vedic deities have become insignificant, but there is considerable speculation about how things came to be what they are. These Upanishads are full of reverence for the natural world, in which trees play an important part:

Truly man is just like a tree. His hairs are the leaves and his skin resembles the natural bark. His blood streams forth out of his skin like the sap of a tree when he is cut . . . The flesh is comparable to wood, the sinews are like the inner bark, the bones are the inner core of the wood and the marrow resembles the pith of the tree.[26]

Of this great tree, if someone should strike at its root, it would bleed but still live. If someone should strike at its middle, it would bleed but still live. If someone should strike at its top it would bleed but still live. Being pervaded by *ātman* it continues to stand, eagerly drinking in moisture and rejoicing. . . . If the life leaves the whole, the whole dries up.[27]

By the time of the later Upanishads the early Vedic deities had been supplemented or replaced by a smaller number with more clearly defined roles. Thus the Śvetāśvatara invokes Īśvara, the Supreme Lord:

He who is the supreme Mighty Lord (*maheśvara*) of lords, who is the highest deity of deities, the supreme master of masters, transcendent, him let us know as God, the lord of the world, the adorable.[28]

Here Īśvara has become the manifested *Brahman* and is omnipotent and omniscient. Souls (*ātman, jīva*) and the world *(jagat, prakṛti)* are united in the one Supreme Reality. The Upanishad echoes the outlook of some of the philosophical and religious schools prevailing at the time of its composition and tries to reconcile them.

The considerable diversity within the Upanishads and Vedic literature generally led to the formation of a variety of exegetical schools. The schools of the Vedānta differed about the relationship between *Brahman* and *ātman*, but agreed that the Upanishads are the authoritative source of knowledge about them. However, the Pūrva Mīmāṁsakas (i.e. earlier commentaries on the Vedas) maintained that the sections of the Vedas concerned with the performance of the sacrificial ritual are primary. According to Julius Lipner:

The Pūrva Mīmāṁsaka emphasis on public religious ritual and its efficacy struck and reinforced an answering chord deep in Hindu minds . . . Though the performance of the solemn Vedic ritual on a large scale may have died down by about the sixteenth century, it was the original Pūrva Mīmāṁsaka concern for ritual that stoked the continuing Hindu liturgical preoccupation with . . . incense and flowers, rites and ceremonies, both in the temple and in the home. Hindus are inveterately ritual-minded – sometimes in an enlightened way, sometimes even to the point of superstition – and the model for ritual, embedded deep in their psyche, is the Vedic *yajña*.[29]

The four Vedas are *śruti*, the most authoritative scripture. All else is remembered and secondary, *smṛti*, though in practice some later literature such as the Bhagavadgītā has come to be regarded by some as equivalent to *śruti*. *Smṛti* is the cumulative tradition through which we hear the voice of *śruti,* and it includes certain Purāṇas (which expatiate upon the youthful exploits of Krishna, among other things), the Rāmāyaṇa and Mahābhārata (which contains the Bhagavadgītā), the six philosophical perspectives known as *darśanas* (Sāṁkhya, Yoga, Nyāya, Vaiśeṣika and the two Mīmāṁsās, Pūrva and Uttara), the Āgamas of the Śaiva Siddhānta and the Vedantic commentaries of Śaṁkara, Rāmānuja and Madhva.

Śaṁkara (788–820 CE) was the founder of the most influential school of Vedānta, known as *advaita* (non-dualism). He interpreted the central *ātman/Brahman* equation of the Chāndogya Upanishad to mean precisely what it appears to mean, namely that *Brahman*, the one reality underlying all diversity, is our true self or *ātman*. Although he distinguishes between an empirical and an ultimate level of reality, it is important to recognize that the former is as real for him as we are ourselves. Lipner underlines this point as follows:

> The common supposition that Śaṁkara taught that the world is an illusion is a much too superficial reading of his thought. For Śaṁkara the world is as real as we are; only the fabric of worldly reality of which we are an integral part has no ultimate reality status . . . Like Hindu thinkers in general, he was careful to distinguish the cognitive scope of scripture from the cognitive scope of empirical experience.[30]

Rāmānuja (eleventh century CE) maintained that *ātman* and *Brahman* are relative, like part and whole, but not identical. *Brahman* is the supreme One and the universal cause who discloses the pre-existent Veda to Brahmā, who reveals it to the Sages. The world is *Brahman*'s 'body'. Brahmā, the creator, lives within and binds together all things as their inner source of life, or *antaryāmin* (a favourite term of Rāmānuja, sometimes translated as 'inner controller').

We must be wary of a tendency to regard *ātman* as some sort of 'ghost in the human machine'. *Ātman* pervades our entire psychophysical being and mental activity is a reflection of the interaction between the two. When we die, both our corporeal body and our centre of mental activity are destroyed, but our *liṅga śarīra* continues until it identifies an impending birth. The *liṅga śarīra* is a *prakṛtic* substrate sometimes called the subtle body, which is not susceptible to normal sense experiences. It contains our memory store and the accumulated karma of previous lives.

Surveys I carried out in the 1970s indicate that many urban, scientifically educated Hindus no longer believe in rebirth.[31] Lipner comments on this tendency, offering a novel ecological alternative explanation:

> A number of Hindus attribute a symbolic significance to the belief in karma and rebirth. Thus, they say, it may not be literally true that rebirth takes place in order to expend accumulated karma, but this is a potent way of symbolizing the responsibilities that one generation of human beings bears in respect of succeeding generations. Current ecological sensitivities give point to this perception. The child is father of the man

in a new sense: are we not reborn in our children who will have to face the consequences of our physical, social and environmental decisions – the 'karma' we have created – in the lives and world that they inherit from us?[32]

We have considered the probable origin of the caste system, but we must also note the manner in which caste and *dharma* (social and cosmic order) are associated via the concept of *varṇāśrama dharma*. *Varṇa* denotes the appearance or form of caste and *varṇa dharma* is the ordered hierarchical *dharma* of caste. *Āśrama* means 'resting place' and refers in this context to the stages of life open to twice-born males (i.e. the three top castes). These are *brahmacarya*, the stage of the student of religious *dharma*; *gārhasthya*, that of the householder; *vānaprastha*, that of the forest-dweller; and *saṁnyāsa*, the stage of the renouncer. Although the majority of Hindus pay little attention to these, they remain a potent reminder of the importance of frugal living in harmony with nature.

The post-Vedic period

The Aryans who migrated from the Punjab to the east not later than 1000 BCE were very different from those who had arrived earlier (*c.* 1750 BCE). They continued to use chariots, horses and cattle, including the distinctive Indus humped cattle depicted on seals. They had the plough and a variety of skills such as pottery, carpentry and weaving. By the seventh century BCE the centre of Aryan domination had shifted from the Punjab to the Gangetic valley where Buddhism would soon begin to exert its influence.

By 600 BCE the Gangetic valley contained a variety of separate social groups in different stages of development. What are now Uttar Pradesh and Bihar probably still contained pre-Aryan tribes, and sections of the forest along their northern perimeters had been 'eaten up' by Agni.[33] Bengal was covered by low-lying swamps and humid forests. Aryan kingdoms vied and sometimes fought with one another for the best territories. The Mahābhārata and Rāmāyaṇa epics reflect the mood of this period. The Rāmāyaṇa, incidentally, maintains that women could perform domestic and certain other rituals in their own right. But even prior to the repressive Laws of Manu (second century BCE or possibly later), a reaction had taken place, with the result that women were no longer allowed to recite the Vedas and had become a source of ritual pollution to men.

Kauṭilya's Arthaśāstra, sections of which were probably written in the fourth to third century BCE, gives an ideal account of life in a well-ordered state. It was composed by an adviser of the Mauryan king Candragupta and reflects among other things the advanced scientific and technical knowledge at that time. Mining, metallurgy, engineering, chemistry, medicine and botany all appear in this compendium, which also includes detailed instructions about the maintenance of forests:

> On non-agricultural land . . . a forest should be made with a single entry. Fruit trees, beautiful groves and attractive flowers should be planted. There should not be any trees of the [thorny] type. A small pond should be there. Deer and similar familiar animals should be there but the teeth and nails of hunting animals must be extracted. Elephants, both male and female, should be there and also children.[34]

> The boundaries of a village shall be denoted by a river, a mountain, forests, bulbous plants, caves . . . or by trees such as *śalmali, śamī* and *kṣīravṛkṣa*.[35]

The first and third of these trees may be what are now known as silk cotton and milk trees; the second is *Acacia suma*. The Arthaśāstra also inveighs against those who pollute public places, temples, ponds and rivers and prescribes corresponding punishments.

The Arthaśāstra is distinctive in that it makes a clear separation between political and religious authority. According to Lipner:

> There was to be no established religion, though this does not mean that in the course of history Hindu rulers did not try to favour or enforce a particular faith, even by means of persecution. On the whole, however, rulers have followed this directive, thus adding to the image of Hindu religious tolerance. Perhaps this tradition helps to explain why independence from colonial rule could be negotiated in terms of a 'secular' state, which in the Indian context simply means that while one has the right to practise a religion, no particular religion is constitutionally privileged. This puts into perspective those Hindu religiously political forces today which seem to wish to act against the weight of history.[36]

An interesting insight into economic changes which occurred between early Vedic times and the period of the Arthaśāstra can be gained from references to cooking methods. Vedic society generally relied on grain and pastoral foodstuffs and only butter fats (*ghee*) were used in ritual. Vegetable oil for cooking was known as *taila*, which is derived from sesamum (*tila*). Although sesamum was known in pre-Aryan times, there are virtually no references to it in the Vedas, whereas the Arthaśāstra mentions the two words more than forty-five times. This may reflect changes from a pastoral to an agrarian economy.

Throughout the period under consideration and well into the early centuries of the Common Era there were interactions between local belief systems and certain well-known Hindu gods. Many spirits were identified with Śiva or Īśvara (which are connected with male phallic worship) and Pārvatī, the goddess of fertility. Other spirits were then associated with them: thus the worship of the spirit of an elephant became the worship of Gaṇeśa, the elephant-headed son of Śiva and Pārvatī. Śiva had already become associated with the pre-Vedic Paśupati, Lord of the Beasts. Śiva himself rides Nandī, the bull, and is festooned with cobras; both these animals signify fertility. The names of the Hindu deities remained the same, but their functions may have changed considerably since their first appearance in religious literature.

Gadgil and Guha have drawn attention to the manner in which such role shifts on the part of the gods may have bearing on the use of natural resources:

> These distinctive local belief systems were now woven together into a composite fabric by identifying many of the spirits with a few key gods in the Hindu pantheon ... This ... had a clear role in regulating and moderating the use of natural resources. It legitimized in a new framework the protection accorded to certain elements of the landscape – for instance groves or ponds near temples, and protection to certain species such as *Ficus religiosa* (peepal) or *Presbytis entellus* (the Hanuman langur) as sacred to a variety of deities. Some of these prescriptions may have been functional in resource conservation, others neutral or even malfunctional. However, identifying them with deities from the Hindu pantheon was an effective way of continuing these practices.[37]

I conclude this historical line of argument for the time being, but will take it up again from the fifth century BCE onwards, when we consider the rise of Buddhism.

A contemporary approach

Although I have emphasized some of the naturalistic aspects of the Vedic tradition – and there are many more – it is important to recognize that these cannot be considered in isolation from the tradition as a whole. It has therefore been necessary, albeit in a very cursory manner, to indicate the major features of the Vedic corpus. Within this totality there has been a tendency on the part of some scholars to overemphasize certain elements (e.g. 'personalist' theism, monism) and virtually ignore others (e.g. panentheism, i.e. the view that the world exists in God).[38] My aim has been to enlarge the framework of the relationship between human communities and their natural surroundings by putting what has often been omitted back into context.

I further illustrate my thesis by summarizing the views of a contemporary Hindu scholar who believes that the Vedas have much to offer the promotion of ecological awareness. Dr Karan Singh is a capable Sanskritist who has made independent translations of the Vedic texts; his publications are readily available in English and he has done a great deal to make successive Indian governments aware of the need for environmental legislation. I offer a brief account of the main tenets of the Hindu tradition as he summarizes them, including his concern for the natural world within this framework.

Karan Singh proposes five doctrines which he believes to be foundational to the Hindu tradition. The first is that underlying the appearance of change and movement represented by *saṁsāra* there exists the unchanging all-pervasive reality which is *Brahman*. The Muṇḍaka Upanishad makes this clear:

> *Brahman* verily is this immortal being.
> In front is *Brahman*, behind is *Brahman*,
> To the right and to the left.
> It spreads forth above and below.
> Verily, *Brahman* is this effulgent universe.[39]

Thus, Karan Singh observes,

> All creation, whether this tiny speck of cosmic dust that we call our world or the billions of galaxies that stretch endlessly into the chasms of time, is in the ultimate analysis a manifestation of the same divine power. Thus, despite the multitudinous manifestations of this space-time continuum, there is ultimately no dichotomy between the human and the divine.[40]

The second major affirmation of the Hindu tradition is that the changing world within human and other living beings is based on the undying *ātman*, which we defined earlier as our true or inner self.

Having established the existence and characteristics of *Brahman* and *ātman*, the Upanishadic seers came to recognize through their spiritual intuition that they are one: 'that self, that *you* are', as Svetaketu's father explains to him in the Chāndogya Upanishad.[41] As we have seen, the exact relationship between the two concepts has been the basis of different Vedāntic schools. This is the third major Hindu tenet.

Karan Singh's fourth basic doctrine is that the supreme goal of life lies in recognizing the existence of the deathless *ātman* within each of us:

> The realization of the *ātman* at once brings an entirely new dimension into the picture, and the realized soul transcends the cycle of suffering, illness, old age and death which are inevitable concomitants of ordinary life, the wheel of change and decay of the manifested universe. [We] may still choose to stay within the limits of manifestations and by [our] presence sweeten the bitter sea of suffering, but [we are] no longer bound to do so.[42]

Finally, there is the imperative to act in a morally appropriate manner for the welfare of all creation. Here Karan Singh utilizes the notion of karma, which on its own means action, in a purposeful and forward-looking manner:

> *Karma* can . . . be considered the moral equivalent of the law of conservation of energy or the equivalence of action and reaction in the field of

natural sciences. While it is true that what we are today is the result of our past deeds, it also follows that we are the makers of our future by the way we act at present. Thus, far from implying fatalism as is often wrongly believed, *karma* gives tremendous responsibility to the individual and places in his own hands the key to his future destiny. Naturally, the unerring law of *karma* can work itself out only over a sufficiently long period of time; therefore the Hindu belief in reincarnation.[43]

Not all Hindus are equally happy with this notion of reincarnation, which Ram Mohan Roy discarded in the nineteenth century. However, within the context of the Bhagavadgītā's understanding of *niṣkāma* karma (i.e. action without attachment), Karan Singh gives it an important interpretation. According to the Gītā, *niṣkāma* karma conventionally means that we must do our work for the benefit of others without considering any benefit for ourselves; rather, we must fix our minds solely on God. Karan Singh maintains that we must enlarge the conventional framework to include the welfare of all creation:

> Finally, there is the concept of welfare, not of any particular person or group or class, but of all creation. As the ancient prayer goes: *Sarve'pi sukhinaḥ santu, sarve santu nirāmayāḥ* (May all beings be happy, may all beings be free from fear). Here welfare is described not in limited terms but as all-embracing, covering not only the human race but also what, in our arrogance, we call 'lower' beings – animals and birds, insects and plants, as well as 'natural' formations, such as mountains and oceans. In addition to the horrors that mankind has perpetrated upon its own members, we have also indulged in a rapacious and ruthless exploitation of the natural environment. Thousands of species have become extinct, millions of acres of forest and other natural habitat laid waste, the land and the air poisoned, the great oceans themselves, the earliest reservoirs of life, polluted beyond belief. And all this has happened because of a limited concept of welfare, an inability to grasp the essential unity of all things, a stubborn refusal to accept the earth not as a material object to be manipulated at will but as a shining, spiritual entity that has over billions of years nurtured consciousness up from the slime of the primeval ocean to where we are today.[44]

Such an interpretation of welfare applicable to the whole of creation is perfectly consistent with the Vedic view of the unity and interrelatedness of all that is – God, selves, the natural world – and although the ancients may not have been as romantically enamoured of nature as they are sometimes made out to be, they undoubtedly believed themselves to be much more a part of it than their present-day successors. Scholars such as Karan Singh may therefore be said to be enlarging the scope of the conventional anthropocentric world view to restore its original cosmic terms of reference rather than modifying the tradition to take on board our contemporary environmental concerns. The distinction is an important one.

———ல்ら———

In this chapter we have considered the early Hindu tradition from an ecological perspective. We have acknowledged the numerous allusions to the beauty and regularity of nature and the parallels that exist between the natural world and the gods, setting these against changes in resource use as many of the earliest hunter-gatherers progressed via nomadic pastoralism towards settled cultivation.

Whether one looks at the development of the Hindu tradition from the perspective of tribals and nomads, the Indus civilization or the Aryans who were responsible for the Vedic texts, considerable closeness exists between humans, the natural world and transcendent Reality (God, the gods, the One, etc.) – so much so, for example, that trees are considered to be 'pervaded by *ātman*'. It is this closeness which encapsulates more than anything else the ecological relationship that once existed between human communities and their surroundings, and which must be rediscovered and combined with the scientific results of researches in a variety of environmental fields if we are to understand past cultures in their totality – to cite the Allchins.

My review of the early Hindu tradition has been extremely sketchy, consisting of a number of ecological 'pointers' attached to a rather thin historical framework. Thus, we have noted the probable significance of Agni 'eating up' the forests and a less convincing explanation of Indra's aggression in terms of the destruction of the irrigation dams of the Indus people. It seems more probable that Indra's thunderbolt (*vajra*) was responsible for releasing the monsoon rains along the southwestern perimeter of the Himalayas.

We noted parallels between the gods of the Indus Valley people and both human and agricultural fertility, and referred to a tribal legend which says that the unecological practices of the Asur iron smelters led to divine retribution, tempered only by the intervention of their womenfolk. I summarized the unecological activities of Arjuna and Krishna in the Khāṇḍava forest in order to be realistic about the shortcomings of certain parts of the Hindu corpus, and challenged the view of some scholars that Śaṁkara taught that the natural world is an illusion.

I concluded the survey with a summary of the views of Karan Singh, a modern scholar whose essentially orthodox exposition of Vedānta acknowledges an ecological dimension of the Vedic scriptures which others have tended to ignore. In so doing we regard him – in Tambiah's sense – as a representative of *continuity* between the past and the present, who is expounding from an ecological perspective what has always been part of historical Vedism. In the next chapter we shall see how other modern Hindus *transformed* the tradition imaginatively.

At the beginning of this chapter we encountered Krishna and Arjuna, racing around the blazing forest of Khāṇḍava in a chariot, killing the wildlife as it tries to escape. In sharp contrast to such ecological irresponsibility, the Bhagavadgītā portrays our heroes in earnest dialogue about the immortality of the soul and of God's universal love. In due course Krishna reveals himself as the *avatāra* of Vishnu:

> As God of all being and established in my creation, I take birth by my spiritual powers. For whenever *dharma* wanes . . . then do I generate an embodied self. For the protection of the good and the destruction of evil-doers, and for the establishing of right I take birth age after age.[45]

Since *dharma* is cosmic in its scope, we would not be forcing the meaning of this passage by claiming that God's love and justice are for creation in its entirety. The Gītā also sets out the path of morally purposeful and selfless action, which is *niṣkāma* karma, selfless work without concern for reward and – as Karan Singh observes – to benefit the whole of creation.

3 Ecology and modern India

In the early 1970s a group of university students in Delhi were asked to assist with the making of a television documentary about Gandhi. The completed film began with the commentator standing on a stage in front of a large cut-out of the Mahatma, explaining what an important film this was going to be. As he left the stage his microphone flex caught around the base of the effigy, which fell flat on its face. The camera swung round to show the auditorium, empty but for two small children sitting on the same chair.

Malcolm Muggeridge made this film at the same time as his better known study of Mother Teresa. Some students at St Stephen's College in the University of Delhi may still think of Gandhi as they did then – backward-looking, village-orientated and politically naïve. Many more, however, would now probably acknowledge the far-sightedness and visionary character of his 'experiments with truth' and his ability to fuse the secular and the religious into a common goal. In particular, the Gandhian notion of *sarvodaya*, with its participative village communities living in harmony with their surroundings, is increasingly acknowledged as preferable to the urban and industrial sprawls of modern India.

However, the contemporary socio-political and ideological milieu of India is unintelligible without some understanding of the forces which shaped the nineteenth and early twentieth centuries, and to these we must now turn.

In the following sections we shall study modern India from the point of view of the changing perceptions that occurred of the relationship between natural resources and society as market forces became dominant. I shall describe scientific and other influences which entered India from the West and were instrumental in causing major social and religious reforms, noting in particular the manner in which these were taken up by reformers such as Swami Vivekananda. Gandhi's philosophical and ethical views will be explored from the point of view of their continuing ecological relevance. Finally, we shall consider a group of outstanding scientists whose researches were shaped by their Hindu beliefs. Among them, Jagadish Chandra Bose investigated the possibility of pain in plants because he believed that from the perspective of Vedānta the boundaries between biology and botany must ultimately disappear. In doing so he effected a transformation of traditional Hindu belief (in Tambiah's sense) which has paved the way for a more ecologically sensitive science than has hitherto been apparent.

Extracting resources

Whereas hunters, nomads and settled cultivators either relied on human muscle and energy from fuelwood, or augmented human power with animal and water power, the industrial mode of energy use both harnessed and mined natural resources for human consumption. It therefore followed an *extractive* pattern of resource use which has depleted natural resources during the past 300 years more severely than anything that preceded them.

The ecological changes that occurred in India during the industrial period were to a large extent a consequence of events in Europe. Technological advances made it possible to turn a wide range of objects into commodities which could be sold for profit. Thus wood, which in pre-industrial times was used primarily for domestic fuel and shelter, could be transformed into paper or burnt as fuel for transportation. As supplies of such raw materials became scarcer and scarcer, the colonizing Europeans looked further and further afield in order to supplement them. By means of the Indian Forest Acts of 1865 and 1878 the British colonial government obtained a monopoly right over valuable areas of forest by declaring them 'reserved forests'. This meant that people who had traditionally enjoyed free access to forests could no longer depend on them for their essential biomass needs. With these changes came the breakdown of cooperatives and local community initiatives, to be replaced by entrepreneurs who exercised considerable monopolies. The market therefore reigned supreme. According to Gadgil and Guha:

> With manufacturing and commerce the dominant activities, markets became the focal point for organizing access to resources. This new belief-system that developed therefore transferred to the institution of the market the veneration reserved for spirits resident in trees by food-gatherers, and in an abstract God by Christian food-producers. Success and status were now clearly measured in terms of money, the currency of the market.[1]

Not only did the Europeans devastate the forests to build ships, to supply sleepers for railways and to manufacture paper and furniture, but they stripped even the most remote regions of India of particular tree species such as teak, *sal* and *deodar* (cedar). Even where forests were eventually replaced, the trees that were planted were mainly those of commercial value. In the Himalayas, for example, mixed forests of conifers and broad-leaved species well suited to filtering and regulating rainwater were replaced with a single strain of commercially valuable conifers. The extractive path of energy usage took an enormous toll of forest ecosystems and progressively marginalized their largely tribal inhabitants. Women, the traditional collectors of fuelwood, were the worst affected:

> The worst sufferers are women who have traditionally been dealing with the resource. Their workload increased . . . Women have to travel the extra distance to collect food, fuel and fodder. Older women and children who used to help the housewife until then are unable to do so any more. Consequently, the housewife is forced to work more than in the past, to collect less food . . . Besides, she is deprived of the medicinal herbs, the allopathic health centres in towns are far from the village, and are open only during the day when she has to be working in the village. Moreover, because of the additional workload, even pregnant women are forced to keep working till a week before childbirth . . . [The] consequence is deterioration both of her nutritional and health status more than that of men.[2]

Not all colonial administrators were insensitive to the need to maintain a balance between forests and their ecosystems. Dietrich Brandis, inspector-general of forests in Madras (now Chennai), advocated the restricted take-over of forests by the state, justifying his view in terms of a mixture of traditional equity and efficiency. He wrote enthusiastically about the sacred groves to be found throughout the sub-continent, describing them as 'the traditional form of forest preservation'.[3]

The industrial mode of energy use in Europe which was responsible for much of the initial large-scale deforestation in India was made possible by technological advances which reached India during a comparatively short period of time around the middle and latter part of the nineteenth century. These were accompanied by a wide range of European notions which were to a large extent shaped by the scientific and rational ideas of the Renaissance. The net result was that India experienced a social and cultural reformation far more extensive and irreversible than anything that had preceded it.

The Reformers

The potency of the ideas that entered India from the West was largely due to the fact that they were presented not as closed, dogmatic systems of thought, but as universal ideas in a secular framework in the English language. They were assimilated by the better educated and more affluent sections of Indian society, especially in the north, and it was from these echelons that most of the significant social, religious and political reformers were drawn.[4]

The concept of secularization is appropriate for encompassing the wide-ranging transitions which occurred during this period and I shall define it, following M. N. Srinivas and others, as the process whereby areas of thought and experience once governed by tradition are progressively determined according to secular criteria.[5] Defined in this manner secularization carries profound implications for religion and society generally, but is not in itself anti-religious. The term should be distinguished from 'secularism', which is opposed to religion.

The nineteenth-century responses to secularization were of three kinds. There were some people who, in the face of an onslaught of scientific and rationalistic propaganda from the West, completely *rejected* their traditional world view. Others *adapted* their heritage by borrowing from some of the new ideas. A third group attempted to show that Indian thought contains all the elements needed to come to terms with the West but that these must be *reasserted*.[6] We shall consider these three categories of response with reference to particular religious thinkers and groups.

The first of the reform movements chronologically was the Brahmo Samaj, which was established in the early 1830s by Ram Mohan Roy (1772–1833). The Samaj was more an agent of secularization than a response to it, and we shall therefore not discuss it in detail. Roy was a forceful advocate of scientific rationalism and social reform, and is often described as the 'father of modern India'. He rejected reincarnation – regarded by most as a cardinal Hindu belief – as being incompatible with science. He was succeeded as leader of the Samaj by Debendranath Tagore (1817–1905) and Keshub Chandra Sen (1838–84). In spite of its influence, the Brahmo Samaj was never large numerically, and by the end of the nineteenth century it had more or less petered out.

The type of response to secularization characterized by the rejection of tradition was scarcely apparent for most of the nineteenth century. In Bengal at the end of the century there were references to the educated, middle-class 'man without a *dharma*' – 'wading his way to liberalism through tumblers of beer'.[7] But there was no organized group of rejecters of religion, and it was not until the twentieth century that intellectual sceptics such as M. N. Roy and Jawaharlal Nehru appeared.

The reasserters of religious tradition were represented primarily by the Arya Samaj, which was founded in 1875 by Dayananda Sarasvati (1824–83). Dayananda believed that the Vedas are the source of all truth and the fountainhead of both science and religion. The proof that they are the sole revelation from God is to be found in their correspondence with nature. He believed that all modern discoveries can be read back into scripture; thus guns, for example, were anticipated by Agni, the god of fire.

In spite of his generally conservative view of scripture, Dayananda was a bold reformer who opposed untouchability, icon worship and child marriage on the grounds that they do not possess Vedic sanction. He also believed that the study of the Vedas should be open to all castes. He was passionately concerned for the welfare of the socially disadvantaged, and his followers set up Arya schools and hospitals, many of which continue today. The movement was particularly influential in northwest India, and in the long term it paved the way for more recent rightist religio-political alignments such as the Bharatiya Janata Party (BJP).

The adapters of religious tradition are best illustrated by the Ramakrishna Mission and its leading exponent, Swami Vivekananda (1862–1902). Sri Ramakrishna (1836–86), in whose name the movement was founded, remained a practitioner of traditional Hindu methods of meditation such as *haṭha yoga* and *tantra* throughout his life. He first encountered the young Vivekananda while he was giving a sermon on the three principles of Vaishnavism: love for God's name, the worship of gurus and universal compassion. Directing his attention to Vivekananda, he said:

> Talk of compassion to beings? Will you, a little animal, bestow compassion on beings? You wretch! Who are you to bestow it? No, no; not compassion to *jīvas* [living beings], but service to them as Śiva.[8]

The basis of Vivekananda's thought is an *advaitic* interpretation of the Upanishads. At the innermost level of being, all is one. This implies, firstly, an egalitarian ethic according to which all differences of race, religion, caste and gender have no ultimate significance – a view Vivekananda held in common with several other reformers. More philosophically, however, it also implies a deeper ontological unity:

> Thinkers in ancient India gradually came to understand that the idea of separateness was erroneous, that there was a connection among all those distinct objects – there was a unity which pervaded the whole universe – trees, shrubs, animals, men, devas, even God Himself; the Advaitin reaching the climax in this line of thought declared all to be but the manifestations of the One. In reality, the metaphysical and the physical universe are one, and the name of this One is Brahman.[9]

Thus, according to Vivekananda, science is the study of the variations which have been manifested by *Brahman*, and since *Brahman* is ultimately one, all branches of science and true knowledge must ultimately converge. This insight was effectively taken up by Jagadish Chandra Bose in his research into the possibility of pain in plants.

Like Dayananda, though less crudely, Vivekananda believed that the basic insights of western thought are anticipated in the Vedic scriptures, but he also utilized various western concepts to develop his own ideas. He tried to show, for example, that the three major schools of Vedānta – *dvaita*, *viśiṣṭādvaita* and *advaita* – are three stages on an evolutionary ladder. However, his concept of 'evolution', for which he utilized the Sāṃkhya term *pariṇāma*, was essentially flawed.[10] He identified *ākāśa* (space) with the ether wind which scientists in the late nineteenth century were trying to detect. He also maintained that our apprehension of ultimate reality as God (Īśvara), living beings (*jīvas*) and the world (*jagat*) is mediated via *māyā*, which he defined as a continuum of space, time and causality.

Vivekananda's interpretation of karma-*yoga* as the basis for this-worldly action is central to his teaching, and paved the way for Gandhian ethics:

> Karma-yoga ... is a system of ethics and religion intended to attain freedom through unselfishness and by good works. The Karma-yogin need not believe in any doctrine whatsoever. He may not ask what his soul is, nor think of any metaphysical speculation. He has got his own special aim of realising selflessness, and he has to work it out himself.[11]

According to Basant Kumar Lal, Vivekananda's interpretation of karma-*yoga* was influenced by the *Gītā*'s concept of *niṣkāma* karma (action without attachment) and, possibly, Buddhism.[12] The following two quotations from the *Complete Works* demonstrate his belief in social action for the sake of others:

> Why should I love everyone? Because they and I are one ... There is this oneness, this solidarity of the whole universe. From the lowest worm that crawls under our feet to the highest beings that ever lived – all have various bodies, but one soul.[13]

> Do you think, so long as one jīva endures in bondage, you will have any liberation? So long as he is not liberated – it may take several lifetimes – you will have to be born to help him.[14]

In common with other nineteenth-century reformers Vivekananda was mainly concerned with the removal of social injustices, many of which appeared to be sanctioned by tradition. Thus, for example, when he alludes to 'the lowest worm that crawls', he is not advocating a primitive form of animal rights, but championing the cause of the most disadvantaged members of society. However, it is easy to see how his affirmation of 'solidarity of the whole universe' can become the basis for an essentially orthodox environmental ethic. Vivekananda's ideas – which continue to be much in evidence in pamphlet form on most Indian university campuses – may therefore be described as ecologically based, though understood in their day only within a limited anthropocentric context.

Indian Muslims and, later, Christians responded to secularization in much the same manner as the Hindu reformers. There were virtually no nineteenth-century rejecters of tradition and the reasserters made their appearance within the political context of nationalism during the twentieth century. The adapters were represented by Sayyid Ahmad Khan (1817–98), who welcomed the rise of science as part of the total growth of human knowledge, and Mohammed Iqbal (1873–1938), who believed that the destiny of humanity is to remake the universe. M. M. Thomas has given a comprehensive account of Christian responses to secularization.[15]

I have made use of the concept of secularization to encompass some of the transitions which took place in India at approximately the same time that the industrial mode of energy use was beginning severely to deplete natural resources. Within the three main categories of response we have noted a general recognition of the importance of the natural world and the need for an affirmative ethic – most convincingly expressed in terms of Vivekananda's exposition of karma-*yoga* – to bring about desirable social reforms. Deforestation and resource depletion were not seen as major problems until the twentieth century, by which time a combination of nationalism and Gandhi's policies were in a better position to address them.

The impact of science

Science was a major component of the secularizing influences which brought about irreversible changes in nineteenth-century India, and it also played an important part in the reformers' restructuring of Hindu and, to some extent, Muslim and Christian thought. Viewed from the perspective of Indian philosophy, some of the scientific discoveries of the West assumed a particular significance, and by the turn of the century a group of outstanding Indian scientists were adopting an approach to their research which displayed a much greater degree of environmental sensitivity than that of their western counterparts. We shall review their work in some detail. First we must consider Darwinism in Europe.

Darwin's view that the animal kingdom had acquired its characteristics over a long period of time was not initially applied to human beings. In *The Origin of Species*, published in 1859, he preferred 'descent with modification' to 'evolution', which had embryological connotations with which he did not agree, and in popular parlance was associated with progress. 'Descent' was a combination of chance (i.e. random variations among offspring) and necessity (i.e. selection in order to survive). The theory was applied to human beings in *The Descent of Man*, published in 1871. Darwin's theory made natural selection the vehicle for 'descent', not, as many preferred to believe, an act of creation and its consequences, or a vital urge or spirit. Thus purpose and direction became superfluous and human beings did not appear to occupy any special place in the world.

The crux of the Victorian reaction to Darwinism was popular revulsion to the notion of common ancestry between humans and animals. This is borne out by the famous debate in Oxford in 1860 between T. H. Huxley and Bishop Wilberforce, and by such outbursts as Archbishop Manning's caricature of Darwinism as 'a brutal philosophy – to wit there is no God and the Ape is our Adam'.[16] Alec Vidler quotes another conservative churchman who pleaded with the Darwinians to 'leave my ancestors in paradise, and I will allow you yours in the Zoological Gardens'.[17]

The reception to Darwinism in Victorian England contrasts sharply with the lack of attention paid to it in India. When *The Origin of Species* first appeared, India was in the throes of the aftermath of the Sepoy uprising. In subsequent years, however, little was said about it in the major journals and there was no controversy comparable to the great furore in England.

This omission is of great significance from the point of view of our understanding of Hindu India and we must therefore document it in more detail. The *Tattvabodhini Patrika*, a popular Bengali monthly founded by Debendranath Tagore, contained a regular column on scientific topics of interest called 'Science News'. Between its inception in 1843 and 1880, there were articles on geology, zoology, physics, chemistry and other branches of science, but no discussion of *The Origin of Species*, though from 1873 onwards illustrated items about anthropology and human evolution had begun to appear.[18] The *Sambad Prabhakar*, a popular but rather conservative Bengali daily founded in 1839, contained a large number of well-informed editorials, often written with a strong bias in favour of science and technology, but nothing significant about Darwinism between 1860 and the end of the century.

These omissions in two of the most important Bengali 'tabloids' are striking. Two other sources in which reactions by educated Indians to Darwinism might have been reflected are the *Proceedings of the Decennial Missionary Conferences* and P. C. Roy's mammoth autobiography, *The Life and Experiences of a Bengali Chemist*.[19] The *Proceedings* contain references to Darwinism in relation to Christianity, but there is no suggestion that Hindu students found it problematic. And P. C. Roy never once mentions Darwinism in the course of 600 pages of descriptive detail of the scientific world in Bengal during the late nineteenth and early twentieth centuries.

The overall impression is that Darwin's views were so readily assimilated by educated Hindus as not to require comment, and that they failed to recognize any threat to their religion comparable to that experienced by the Victorians. If the crux of the matter in England was the common ancestry presupposed by Darwinism between human beings and animals, it is not difficult to understand why educated Hindus reacted differently, since for them reincarnation and a general tendency of gods to assume animal forms took such a belief for granted.

If Darwinism raised fundamental questions about the Victorian understanding of our relationship to other life forms, and the work of archaeologists and geologists greatly extended the time span of the universe, advances in astronomy did at least as much to expand what was known about the size of the universe. In 1832 the astronomer Henderson proved that the nearest star is 24 billion miles away and by the end of the century astrophysicists were paving the way for Einstein's theories of relativity. The coming together of so many hitherto unrelated branches of science under common theories was to many educated Hindus evidence of the truth of Vedānta: the many becoming 'One'.

Indian scientists

From the point of view of India's new generation of scientists the progress of science in the West seemed to demonstrate a Hindu truth: the fundamental unity of all existence. Thus when Indian science became sufficiently advanced for experimental research to be undertaken, these scientists often chose to work on the interface between established scientific disciplines where they believed that the boundaries would eventually disappear. Jagadish Chandra Bose (1858–1937) studied the phenomenon of consciousness in plants because he believed that 'In the multiplicity of phenomena, we should never miss their underlying unity.'[20]

The rediscovery of ancient Indian science became important for Indian scientists at the turn of the nineteenth century and culminated in the publication of the *History of Hindu Chemistry*, in two parts, by P. C. Roy, and *The Positive Sciences of the Ancient Hindus,* by B. N. Seal.[21] The latter, who was professor of mental and moral science at Calcutta University from 1913 to 1920, attempted to interpret classical Sāṁkhya philosophy in terms of Darwinian evolutionary theory. But his use of *pariṇāma* for evolution was not what Darwin meant by it, and his identification of *ākāśa* (space) with the ether 'wind' did not do justice to the complexities of either concept.

Some scientists, however, were much more profound in their attempts to interpret their scientific insights from the standpoint of Hindu philosophy. The Bengali chemist, P. C. Roy (1861–1944), for example, describes his discoveries in terms which echo the 'intuition' or 'integral insight' whereby the one and the many are perceived to be the same. He attributed the combination of rationalism and intuition which he brought to bear on his scientific research to the influence of the Brahmo Samaj.[22] His 'intuitive' discovery of mercurous nitrite in 1895 paved the way for a series of important advances in chemistry.

Satyendra Nath Bose (1894–1974) first made his mark as a physicist by discovering the type of statistics which governs the behaviour of light particles or photons. He worked with Madame Curie in Paris in 1924 and then joined Einstein in his quest for a unified field theory. He gave his name to the boson, an elementary particle which obeys Bose–Einstein statistics, and in 1953 was able to solve some of the equations governing the relationship between electromagnetism and gravitation. Megnad Saha (1893–1956) followed in Bose's footsteps by publishing an original paper on relativity, but his main achievement was the discovery that in a very hot star electrons may be stripped from their parent nucleus to form an electronic 'gas', the particles of which obey Maxwell–Boltzmann statistics like ordinary molecules. In later life he applied his stellar theories to atmospheric ionization beneath the ozone layer.

Srinivasa Ramanujan (1887–1920) never showed any inclination to move from mathematics to some less abstract or more inter-disciplinary area of research, but he did outstanding research on the theory of numbers. Nehru saw in his short but brilliant career a microcosm of India's destiny. C. V.

Raman, K. S. Krishnan, P. C. Mahalanobis, H. J. Bhabha, S. Chandrasekhar and a host of others made outstanding contributions to science. All were proud of their religious heritage (Homi Bhabha, who pioneered India's nuclear programme, was a Parsi), and they tended towards pure rather than applied science. However, it was Jagadish Chandra Bose more than any of his contemporaries who permitted his choice of research to be influenced by religious convictions, and we shall therefore consider his work in more detail.

Jagadish Chandra Bose

Jagadish Chandra Bose began his research career by clarifying certain aspects of Maxwell's work on electromagnetism. He then conducted a series of experiments designed to study parallels between the behaviour of metal 'coherers' and plant and animal responses to stimuli. In 1917 he wrote:

> In the pursuit of my investigation, I was unconsciously led into the border regions of physics and physiology and was amazed to find boundary lines vanishing and points of contact emerge between the realms of the Living and the Non-living . . . A universal reaction seemed to bring together metal, plant and animal under a common law.[23]

On the occasion of the inauguration of the Bose Institute in Calcutta in the same year, he set his quest for a wider synthesis between the sciences against the background of what he described as the excessive specialization of scientists in the West, and claimed that India is well suited by her past to provide a corrective:

> The excessive specialization in modern science has led to the danger of losing sight of the fundamental fact that there can be one truth, one science which includes all branches of knowledge. How chaotic appear the happenings in Nature! Is Nature a cosmos in which [the] human mind is some day to realize the uniform march of sequence, order and law? India through her habit of mind is peculiarly fitted to realize the idea of unity, and to see in the phenomenal world an orderly universe. It was this trend of thought that led me unconsciously to the dividing frontiers of the different sciences and shaped the courses of my work in its constant alternation between the theoretical and the practical, from the investigation of the organic world to that of organized life and its multifarious activities of growth, of movement, and even of sensation. Thus the lines of physics, of physiology, and of psychology converge and meet.[24]

Bose's interest in botany and physiology was in part a reversion to an early childhood passion and there was also an element of reaction against the West's emphasis upon specialization. He believed, however, that by concentrating his attention upon the boundary areas between the different physical and biological sciences he would ultimately demonstrate the underlying unity of all things. The scientist's quest was thus virtually a religious activity, the ultimate goal of which was the discovery of the 'One' in the many. The following quotations describe the 'theological' bias of the scientist in his desire to question and understand, and his 'moment of truth':

> In my scientific research . . . an unconscious theological bias was also present . . . It is forgotten that He, who surrounded us with this ever-evolving mystery of creation, the ineffable wonder that lies hidden in the microcosm of the dust particle, enclosing within the intricacies of its atomic form all the mystery of the cosmos, had also implanted in us the desire to question and understand.[25]

Bose's language was strongly coloured by Sāṁkhya philosophy – 'microcosm' and 'ever-evolving mystery of creation', for example. The ultimate aim of scientific discovery, however, was the realization of the 'One' in the many:

> When I came upon the mute witness of these self-made records and perceived in them one phase of a pervading unity that bears within it all things: the mote that quivers in ripples of life, the teeming life upon our earth, and the radiant suns that shine above us – it was then that I understood for the first time a little of that message proclaimed by my ancestors on the banks of the Ganges thirty centuries ago: 'They who see but one in all the changing manifestations of this universe, unto them belongs Eternal Truth – unto none else.'[26]

Thus the work of the scientist, properly conducted, is a religious quest whereby we are drawn naturally to search for the wonder that is at the heart of all existence. A corollary of this is that if scientists wish to make any significant progress, they should approach their work with a degree of sensitivity and reverence which was not particularly characteristic of the methodology of western scientists at the end of the nineteenth century. Bose believed that the true scientist must learn to evoke, look and listen, rather than probe and analyse from a distance. After discovering a resonant and oscillating recorder which measured the electrical response of plants to external stimuli, he commented:

> It has been beautifully said – and it is a law of the moral world as unchangeable as physical laws: 'Ask, and it shall be given you; seek and ye shall find; knock, and it shall be opened to you.'[27]

Although Bose disliked what he believed to be the insensitive western attitude to the natural world, however, his experimental methods were acceptable in Europe, and the results of his early botanical work were widely acclaimed. Like P. C. Roy he believed in the importance of intuition, and his attempts to prove the existence of consciousness in lower forms of life were essentially intuitive.

Bose's earliest researches were prompted by the discovery that an electric wave receiver seems to show signs of fatigue after constant use. He then began to wonder to what extent the responses of inert and living matter to external stimuli can be compared. At this point he did research on the interface between plants and animals (i.e. essentially at the level of plasmodesmata and gap junctions, though he did not use these terms). Towards the end of his life he began to investigate whether or not plants possess latent consciousness. The 1945 edition of the *Encyclopaedia Britannica*, published eight years after his death, gives a fair estimate of his work: his research 'was so much in advance of his time that precise evaluation is not possible'.

The Hindu beliefs of Jagadish Chandra Bose are similar to those of Vivekananda since they represent an adaptive response to secularization. However, they go much further, for Bose was able to familiarize himself with the most advanced developments in western science and then use his own indigenous belief system to focus and advance his experimental work. His appreciation of the fact that western science, however successful in many spheres, is not value free, and can therefore, in principle at least, be reshaped along lines which are more environmentally sensitive and compatible with long-term sustainability, represents a major step forward. In Tambiah's sense, Bose was able to achieve a remarkable transformation of Hindu tradition, the consequences of which have yet to be fully appreciated.

Gandhian ethics

We have noted the prominence given at the 1992 Earth Summit to Gandhi's view that 'the earth has enough for everyone's need but not for anyone's greed'. Gandhi's ideas are very often cited as the inspiration behind modern environmental movements. How legitimate are such claims and to what extent do Gandhian thought and action provide a basis for environmental ethics?

Mohandas Karamchand Gandhi (1869–1948) was born in Gujarat. His early Vaishnava upbringing implanted in him a strong sense of God's presence, 'more real than the five senses can ever produce'.[28] His apprehension of God was expressed through feelings, by living a life of truth and purity and in the struggle against evil:

> There is an indefinable mysterious Power that pervades everything. I feel it, though I do not see it. It is the unseen Power which makes itself felt and yet defies all proof, because it is so unlike all that I perceive through my senses. It transcends the senses.[29]

> I know too that I shall never know God if I do not wrestle with and against evil even at the cost of life itself. I am fortified in the belief by my

own humble and limited experience. The purer I try to become, the nearer I feel to be to God.[30]

The second statement was based on Gandhi's mature reflections, but the intensity of his feeling for God stemmed from the early formative Vaishnava influence. He could also say, 'I believe in *advaita*. I believe in the essential unity of man, and for that matter of all that lives.'[31] Such a creed carries an implicit ethic which recognizes the intrinsic value of all people and all life.

Gandhi was also influenced in his youth by the Jain notion of *ahiṁsā*. This is traditionally understood by Jains to mean non-injury to all beings, in thought, speech and action. Gandhi was less strict in his interpretation: mosquitoes could be destroyed and, in certain circumstances, even human life. He also attributed positive characteristics to *ahiṁsā*, such as the removal of anger, malice, jealousy and all else that stands in the way of love. Far from being a passive attribute, Gandhi sometimes described it as soul-*force*.

Gandhi married at an early age and went to England to study law. There he became acquainted with Christianity and also studied the Bhagavadgītā, both of which made a deep impression on him. In 1891 he returned to India, where he practised somewhat unsuccessfully as a lawyer. Two years later he went to South Africa to assist an Indian merchant with a lawsuit.

In South Africa Gandhi became the first Indian barrister in Natal. He espoused the rights of Indian labourers and founded the Natal Indian Congress. During a brief visit to India he met G. K. Gokhale (1866–1915), who made a lasting impression on him, directing his attention to important issues such as the plight of untouchables and the need for Hindu–Muslim unity. He records in his *Autobiography* that by 1903 he was beginning to explore the ethical teaching of the Gītā from the perspective of his legal training:

> I understood more clearly in the light of the Gītā teaching the implication of the word 'trustee'. My regard for jurisprudence increased, I discovered in it religion. I understood the Gītā teaching of non-possession to mean that those who desired salvation should act like the trustee who, though having control over great possessions, regards not an iota of them as his own. It became clear to me as daylight that non-possession and equability pre-supposed a change of heart, a change of attitude.[32]

This is essentially the doctrine of *niṣkāma* karma, which formed the basis of Vivekananda's interpretation of karma-*yoga*.

In 1904 Gandhi read Ruskin's *Unto This Last*, which so impressed him that he later translated it into Gujarati under the title *Sarvodaya*:

> The teachings of *Unto This Last* I understood to be:
> 1   That the good of the individual is contained in the good of all.

2   That a lawyer's work has the same value as the barber's, inasmuch as all have the same right of earning their livelihood from their work.
3   That a life of labour, i.e. the life of the tiller of the soil and the handicraftsman, is the life worth living.

The first of these I knew. The second I had dimly realized. The third never occurred to me. *Unto This Last* made it as clear as daylight for me that the second and the third were contained in the first.[33]

Sarvodaya means the awakening of all, and its political expression is in the decentralization of power. Gandhi believed that opportunities for personal initiatives are maximized in the *panchayat* system, consisting of voluntary cooperation among members of self-contained villages based primarily on agriculture and cottage industries:

> In this structure composed of innumerable villages . . . life will not be a pyramid with the apex sustained by the bottom. But it will be an oceanic circle whose centre will be the individual . . . the outermost circumference will not wield power to crush the inner circle, but it will give strength to all within and derive its own strength from it.[34]

Such participative village republics are much more conducive to a sustainable relationship between people and their natural environment than urban-industrial societies.

Ruskin's ideas helped Gandhi to appreciate the significance of manual labour as an expression of solidarity between the educated and uneducated, and encouraged him to set up a rural settlement for his followers. Phoenix Farm was his first ashram.

In 1906 the Transvaal Government introduced the Asiatic Registration Bill, which was designed to restrict the movement of Indians. Gandhi opposed it and was put in jail. At about this time he first referred to *satyāgraha*, which means 'laying hold of truth or reality'. He believed that as truth becomes part of one's life in a total manner so one's ability to apprehend it is enlarged and ultimately seeks the well-being of the whole world. *Satyāgraha* therefore becomes the vehicle for *ahiṁsā*, the scope of which is all-embracing:

> To see the universal and all pervading Spirit of Truth face to face one must be able to love the meanest of creation as oneself.[35]

Tolstoy Farm was set up in 1910 to cater for *satyāgrahis* and their families. Gandhi's manner of administration was a strange blend of anarchy and personal authoritarianism. Young men and women could live and even bathe together, but when the first 'lapse' occurred he had all the girl's hair cut off to render her 'sexless', while he himself fasted for seven days as a penance. The young man got off scot-free!

At about this time Gandhi took a vow involving the three basic principles of *brahmacarya* (celibacy), *ahiṁsā* and *satyāgraha*. He taught these principles at Tolstoy Farm. In 1914 he returned to India where he established Sabarmati Ashram near Ahmedabad. There he taught the same basic principles, but also expected the members to spin and weave. They were to possess nothing, boycott imported cloth and accept untouchables as their equal (at that time about one-fifth of India's population were untouchables).

Gandhi's political philosophy in relation to growing Indian nationalism and the excesses of colonialism has been evaluated by Bhikhu Parekh.[36] He appreciated much of what had originated in the West and resisted much more. Thus reason and rationalists were 'admirable beings', but when rationalism claimed omnipotence for itself, it became a 'hideous monster'. Asked on one occasion what he thought of western civilization, Gandhi quipped 'I think it would be a good idea.'

Consistent with his understanding of *sarvodaya*, Gandhi believed that only a federally constructed polity based on self-governing local communities was fully democratic, and he used the term *swaraj* to denote true democracy. *Swadeshi* denoted 'belonging to one's own country' in a total sense, that included the need to restore the broken relationship with the natural world, a theme also taken up in Gandhi's spiritual socialism (later called the Constructive Programme). Bhikhu Parekh sees in Gandhi's reinterpretation of the Hindu tradition a rejection of western anthropocentrism in favour of a cosmocentric anthropology:

> He [Gandhi] challenges the anthropocentric view that man enjoys absolute ontological superiority to and the consequent right of unrestrained domination over the non-human world. He rightly argues that the grounds on which such claims rest are philosophically suspect and that the havoc they cause ultimately rebounds on man himself . . . Gandhi's cosmocentric anthropology restores his ontological roots, establishes a more balanced and respectful relationship between him and the natural world, assigns the animals their due place and provides the basis of a more satisfactory and ecologically conscious philosophical anthropology.[37]

Parekh's summary reflects the overall thrust of Gandhi's philosophical outlook, but it is not so easy to give a comparable account of his ethics. I have attempted to summarize the development of his ideas. Thus a modified version of Jain *ahiṁsā* was reinterpreted to express positive well-being to all living beings, a view reinforced by *advaitic* insistence on their ultimate unity. However, that unity is specious unless it is organic and participative, hence the need for *sarvodaya* and *swaraj*. *Satyāgraha*, the practical laying hold of reality (or truth), is the vehicle for *ahiṁsā*; both embrace all life and especially the 'meanest of creation'. *Satyāgraha* builds on Vivekananda's exposition of karma-*yoga*, which in turn incorporates the Gītā's insistence on non-attachment, *niṣkāma* karma, so that we become responsible custodians and trustees (Gandhi's legal term) of our planet.

We have seen how the industrial mode of energy use led to an unprecedented level of deforestation and the destruction of communities which had depended on forests for their well-being. Parallel to these developments and related to them in terms of scientific and technological advances was the large-scale exposure of India to western notions, which entered the subcontinent as universal ideas in a secular framework in the English language.

The resulting secularization created a ferment which ultimately led to a range of social reforms and the restructuring of traditional beliefs. Three major categories of response occurred whereby traditional views were rejected, reasserted or adapted; these were illustrated with reference to the reform movements and the beliefs of individual thinkers during the second half of the nineteenth century. Among them was Swami Vivekananda, whose insistence on 'the solidarity of the whole universe', ranging from 'the lowest worm that crawls . . . to the highest beings that ever lived', might have formed the basis for an environmental ethic, had his main concern not been the removal of social inequality. His affirmative this-worldly ethic, couched in terms of the karma-*yoga* of the Bhagavadgītā, exerted a strong influence on Gandhi.

Although science was responsible for much of the questioning of traditional values, it was also used imaginatively by several reformers to restructure their beliefs. Darwinism, which was heralded with suspicion in Victorian England because of the common ancestry it presupposed between humans and animals, was accepted by educated Hindus without any such difficulties, since for them humans and animals were considered to be much closer than in the West.

Indian scientists were particularly interested in the manner in which heat, light, sound, electricity and magnetism, which were separate disciplines in their own right at the beginning of the century, had become united under common scientific theories by the end, culminating in Einstein's quest for a unified field theory. They interpreted this as the fulfilment of Vedānta, whereby the unity underlying diverse phenomena was becoming apparent. This prompted several of them to select research topics on the borderlines of hitherto separate scientific disciplines, and to investigate, for example, the possibility of pain in plants.

I described in detail the work of Jagadish Chandra Bose, whose brilliant researches in the natural and biological sciences demonstrated how Indian philosophical thought can be brought to bear on the most advanced problems of western science, noting in particular the implications of his work for the need for more environmentally sensitive science.[38] In keeping with Tambiah's emphasis on the 'continuities and transformations' between the past and the present, we regard Bose's work as an imaginative 'transformation'.

The ethical implications of the view that unity underlies all diversity were explored by M. K. Gandhi, who developed Vivekananda's exposition of karma-*yoga* to address secular social and political issues from within a spiritual framework – 'an unparalleled achievement in the history of Hindu thought', to quote Bhikhu Parekh.[39] Gandhi also reinterpreted *ahiṁsā* as the irresistible force of love, and by his personal example created *satyāgraha*, the laying hold of reality or truth, as its vehicle. In these and related ideas such as *swaraj, swadeshi* and *sarvodaya,* Gandhi challenged the western anthropocentric view that human beings possess the right of unrestrained domination over the natural world, replacing it with a cosmocentric anthropology based on mutually enriching relationships between humanity and nature.

4 Struggles for the forests

The journey over these winding mountain roads had tired them. One or two had even opened a bottle of the local liquor on the way. Others lolled on the ground eating their food, watching the women warily. A strange silence enveloped the forest.

Shankarbabu broke the silence. He stumbled forward drunkenly, brandishing a gun and yelling, 'Go away, you foolish low-caste women, or else the police will arrest you. We've paid good money to cut the trees, we have a permit. Go away.'

The sight of the gun frightened some of the women, but not Gaura Devi. She stood her ground, looked the drunkard straight in the eye and challenged him, 'Shoot and only then can you cut this forest which is like a mother to us. Its herbs take care of our children when they are ill, its roots build the earth and prevent the soil from slipping away, its strong branches support our houses, our very lives! Shoot, you coward!'

Her determined, powerful voice unnerved Shankarbabu . . .[1]

It was by means of such encounters, told and retold in music and drama, and recently published as children's schoolbooks, that the Chipko and Appiko movements gathered momentum to such an extent that the large-scale destruction of forests was slowed down, and in some cases halted. The Chipko movement began in the lower regions of the Himalayas to the north of Uttar Pradesh, and we must shift our focus to this area.

In this chapter we shall consider the reasons why peasant communities opposed the commercial ambitions of both colonial and post-colonial governments with such determination, how the Chipko movement grew out of earlier struggles, and the emergence of leaders such as Sunderlal Bahuguna and Chandi Prasad Bhatt. It will be apparent that although these charismatic personalities had little direct contact with Mahatma Gandhi, they readily acknowledged the influence of his ideas and methods. Bahuguna also acknowledged the influence of Jagadish Chandra Bose.[2] We shall consider another important contemporary figure, Anna Hazare, whose primary contribution to eco-restoration lies in the field of water management in rural Maharashtra and who acknowledges his indebtedness to Swami Vivekananda.[3]

Many of the environmental activists with whom this chapter is concerned were building on traditional Hindu views about the value of all life and the need to live in harmony with nature. Much of what we shall describe therefore represents continuity with the past. There are, however, certain instances where the past is imaginatively transformed, in Tambiah's sense. The embracing of trees in order to protect them from felling may be cited as one example (though this had its antecedents in the activities of the eighteenth-century Bishnois). Chandi Prasad Bhatt's interest in appropriate technology is another, while Sunderlal Bahuguna's use of pilgrimage (*padayātrā*) represents a potent transformation of traditional belief in the mystical properties of the land.

The Himalayas

The Himalayan region of India covers an area of 506,000 square kilometres and contains a predominantly rural population of 21 million. It comprises the states of Jammu and Kashmir, Himachel Pradesh, Uttarakhand (the Garhwal and Kumaun hills) in Uttar Pradesh, the Darjeeling district of West Bengal, Sikkim and the northeastern states of Arunachal Pradesh, Nagaland, Mizoram, Manipur and Meghalaya. Nepal and Bhutan are independent kingdoms. The entire mountain chain stretches a distance of 2,500 kilometres.

The snowbound peaks of the high Himalayas feed perennial rivers which make the northern part of India extremely prosperous agriculturally and, increasingly, in terms of hydro-electricity. The forests play a major part in maintaining the hydrological and nutrient cycles and the climate balance. Broad-leaved trees such as *banj* (oak) slow down rain run-off and control soil erosion and landslides by reducing the impact of rainwater through their canopy cover and by absorbing water in their roots and fallen leaves. Leaf litter provides organic nutrients to agricultural land and fodder for livestock. Forests are home to a vast number of species of fauna and flora.

The Himalayas contain a variety of moist and dry temperate forests which give way to alpine vegetation at high altitudes. Although these forests originally contained a mixture of tree types, there is today a progression from *sal* and *chir* pine in the lower hills through cedar and oak to cypress and rhododendron at the upper tree line. Vegetation in the central Himalayas ceases at about 4,000 metres.

Our main concern is with the central Himalayan region of Uttarakhand, which includes the eight hill districts of northern Uttar Pradesh. These are Pauri, Chamoli, Tehri, Uttarkashi and Dehra Dun (collectively known as Garhwal) and Almora, Pithoragarh and Nainital (collectively known as Kumaun). These are the areas in which the first protests against commercial logging occurred and where the Chipko movement was born.

Tehri Garhwal was a princely state whose ruler was head of the famous Badrinath temple; his lineage stretched back many centuries. Kumaun was a colonial territory with a less feudal and more bureaucratic administration. Thus Uttarakhand was divided into two very different political systems. The dominant castes in both were brahmins and *rajputs* who had migrated into the hills from the plains, possibly, initially at least, to visit the hill temples.

Sunderlal Bahuguna endorses the accounts by various historians of the traditional prosperity of the people of Garhwal and Kumaun:

> The peasants of Garhwal and Kumaun are better off than the peasants in any part of the world, who neither live in such well-built houses, nor are so well-dressed as the peasants of Kumaun . . . The people . . . used to export wheat, rice, coarse grains, oil seeds, ginger, saffron, herbs, walnut, handmade paper, copper rods, musk, honey, *ghee*, woollen clothes, cows, bulls, ponies etc., in the markets of the foothills and imported only *gur* (molasses) and cotton cloth . . . The highlanders have been a relatively prosperous people and the source of their prosperity was the natural forest managed by village communities. There were mixed forests – conifers with broad-leaved species providing wild fruits, roots, honey, vegetables, fodder and fibre.[4]

These hill societies possessed a strong sense of cohesion and there is little evidence of class divisions within them. Village *panchayats* dealt with judicial as well as social and religious matters even during colonial rule. Women worked equally with men in cultivation and in everything except ploughing, while also assuming domestic responsibility at home.

The majority of villages in Garhwal and Kumaun were situated below oak forests so they benefited from the filtered water released by them and above the fields along the course of a river. Situated between about 1,000 metres and 2,000 metres above sea level, they had easy access to forests where cattle could graze and from which fuel and agricultural implements could be obtained. The lower *chir* forests were also used for pasture, and every year dry grass and pine-needle litter in them was burnt in preparation for the next crop of grass.

The dependence of the hill people on the forests was reflected in a variety of religious customs. Hilltops were dedicated to gods and surrounded by sacred groves usually of cedar, though birch could also be used. The planting of a sacred grove was regarded as 'a work of great religious merit'.[5] In Tehri leaves were offered to Patna Devi, the goddess of leaves.

Rules also existed among villagers regulating the extent to which branches could be cut from certain trees during the year. The amount of grass that could be cut by each family was assessed and new plants and saplings were carefully guarded by villagers. Robert Redfield has summed up these traditional management practices by ascribing to the Uttarakhand peasantry an 'intimate and reverential attitude toward the land'.[6]

Colonial forestry

The first attempt to exploit India's forests for commercial gain occurred in 1796 when a syndicate was set up in Malabar to extract teak for shipbuilding and military purposes. Early in the nineteenth century as the East India Company consolidated its military hold over new territories, it became standard procedure to raze forests to the ground as a symbol of victory. Following a decision in the middle of the century to expand the railways, vast areas of *sal* and teak forest were felled, and as these became progressively exhausted the hill areas of the lower Himalayas were scoured for cedar.

The first colonial Forest Act was passed in 1865. In 1878 a more comprehensive Act closed or reserved certain forests so they could be used exclusively to produce timber. The Act provided for three types of forest: reserved forests for sustained exploitation, where state control was total and peasants' rights were either denied, transferred or permitted to a limited extent; protected forests, also controlled by the state, where commercially valuable trees were earmarked for cutting and which could be closed periodically; and *panchayat* forests controlled by village communities. These were very few in number. The 1878 decision to annex virtually all forests was the outcome of a lengthy series of discussions which have been described by Gadgil and Guha as an example of an early environmental debate.[7]

Even prior to the 1878 Act the magnificent cedar and pine forests of Garhwal had been leased by the king to the colonial rulers. In many areas villagers were progressively denied their traditional rights and the forests were managed for profit by the Imperial Forest Department. The cedar forests of Himachal Pradesh and the *sal* forests of Sikkim were similarly leased by their rulers. The peasantry in Garhwal were granted certain concessions by the king (e.g. they could collect fallen branches) in return for their assistance in putting out fires.

In Kumaun the colonial government began its onslaught on the forests early in the nineteenth century by cutting *sal* trees in the foothills. Between the 1878 Act and about 1890 many forests were declared reserved. Then in 1893 waste land which was neither reserved nor fell within the perimeter of villages was declared district protected forest, which severely restricted access by villagers. This territory included lakes, rivers, temple lands, pasture grounds and roads. At about the same time government interest in *chir* pines was increased by the discovery that they could be tapped to produce commercially useful turpentine and resin.

It was government policy wherever possible to transform mixed forests of broad-leaved trees and conifers into monocultural areas of commercially valuable trees – *sal*, teak and *deodar* (cedar). Sunderlal Bahuguna describes the effects of such substitution of species as follows:

> The change in tree species had a disastrous effect on local ecology and economy. The pressure of grazing and meeting the other requirements of

the local communities fell upon the civil and community forests which disappeared in no time. Water sources dried up, erosion was accelerated and the process of fertile soil manufacture in the forests was greatly hindered, because forests are acidic and the rate of decomposition of pine leaf litter is considerably lower than that of other species. A decline in the fertility of land and an increase in population resulted in further deforestation. All over the Himalayas hungry people have encroached upon forest land.[8]

The peasants of Uttarakhand did not cooperate with the removal of their traditional rights and there were many violations of government restrictions. From their point of view the three most contentious areas were grazing, the cutting of branches and the burning of the forest floor. The government eventually modified its stance to allow grazing in the vicinity of inflammable brushwood, and branch cutting to take place only on broad-leaved species of trees deemed to be of no commercial value. A compromise was also reached about when and where forest floor burning could take place. However, the 1914–18 War provided the government with reasons to bear down even more heavily on India's forests.

The colonial justification for many of its policies was summed up by the concept of 'scientific forestry', which Gadgil and Guha have convincingly demonstrated to be a complete misnomer:

> [European] science and technology had explicitly stated assumptions about the working of nature: it recognized only those natural forces whose working could be observed empirically . . . The same methodology was then sought to be applied to more complex phenomena, including the working of ecosystems and human societies. There were serious difficulties . . . Consider, for example, the functioning of a tropical humid forest, which is an interacting system of hundreds of plant species, along with thousands of species of micro-organisms as well as animals of varying sizes. To this day we do not even have a complete inventory of all the living organisms from a single hectare of any one tropical humid forest . . . How can one then apply this supposedly 'scientific' method in decisions on how to manipulate a resource base such as a tropical humid forest? Obviously, all that is possible are a few crude prescriptions based on rules of thumb. If the object was the conservation of resources, these rules would differ scarcely at all from pre-scientific prescriptions of earlier times . . . In our 'scientific' system sacred groves may be replaced by preservation plots, but the notion is the same, and the establishment of preservation plots has exactly the same arbitrariness.[9]

The colonial claim for scientific legitimacy with regard to its forest policies was flawed in terms of western scientific methodology because it did not presuppose any model based on empirical data capable of being modified in the light of its predictions. And at a deeper level, western science at that time lacked the ability to probe the interdisciplinary boundaries and relationships which were so significant to Jagadish Chandra Bose and his contemporaries. 'Scientific' forestry undermined the community ownership of forests and destroyed the link between people and their natural habitats. The traditional community pattern of forest use may not have been as ideal as is sometimes claimed, but the colonial forest policies were ecologically disastrous.

Early opposition

Early in the eighteenth century near Jodhpur in Rajasthan a devout Hindu called Jumba Ji Maharaj established a small sect called the Bishnois ('Followers of the twenty-nine rules'). He and his followers, many of them women, were so devoted to trees that when a local prince ordered the felling of *khejri* (acacia) trees to fuel his lime kilns, they hugged the trees to protect them. Many were killed, but the trees were saved. A temple has been built in honour of the Bishnois martyrs.

This episode differs from later instances of resistance to unjust forest laws in that it was essentially an attempt to protect trees having a particular religious significance and practical usefulness, whereas the struggles against colonial forestry were to prevent the destruction of a total way of life. Between 1831 and 1837 there was an uprising against British administration of forests in Uttara Kannada. The resistance was non-violent, but it was crushed with the help of the army. In 1879 there was a rebellion against new regulations about forests and liquor in the Rampa hills of Andhra Pradesh. It lasted more than a year and spread to other areas, but was eventually put down.

In Tehri Garhwal (the princely part of Garhwal) peasant protest took the form of an appeal to the king to restore justice on account of malpractices by corrupt officials. Such a protest was known as a *dhandak* and was non-violent. The peasants would gather at a pre-arranged meeting place, usually a temple, and would collectively decide not to comply with certain rules. They would then march to the palace to meet the king who, as the embodiment of Badrinath, could be relied upon to redress their grievances. Ramachandra Guha describes the role of the Garhwal king in such matters:

> In general, appeals to the monarch rested on two core assumptions: that the king symbolized the spirit of collectivity and that, as the temporal and spiritual head, he was the fount of justice . . . Traditional societies experience life as an ever-expanding web of connections which reaches beyond local and national communities into the depths of nature. It is the function of the monarch to maintain the harmony of this integration between society and nature, between the microcosmos of human beings and the macrocosmos of gods. As the mediating link between the sacred

and the profane, the king takes on some of the attributes of the gods; he is quasi-divine.[10]

By the end of the nineteenth century the Garhwal king had introduced forest regulations in line with those of the colonial state in neighbouring Kumaun. The *dhandaks* became more violent, though there seem to have been few, if any, fatalities. In 1907 there was a mass meeting in Almora (Kumaun) to protest against the policy of declaring all waste lands reserved forests. In 1916, after several unsuccessful attempts to negotiate, a number of government forests and resin depots were set on fire.

Early in 1921 a mass meeting at the annual temple fair at Bageshwar decided to refuse the voluntary service of villagers, known as *utar*, to government officials. A crowd of more than 10,000 heard a message passed on from Mahatma Gandhi that 'he would come and save them from oppression', and it was agreed that they would agitate for greater access to their forests.[11] The summer of 1921 was exceptionally dry, and when the exasperated villagers set fire to the forests, considerable areas were destroyed. Anti-government feelings ran high and some of the popular propaganda compared the colonial power to Rāvaṇa, the evil king of Lanka in the Rāmāyaṇa. The Forest Act of 1927 reduced access to fuel and fodder for survival, while increasing timber productivity.

Further attempts at negotiations led to the reorganization of the reserved forest areas – which had been enlarged to include temple trees and sacred groves of oak – into two classes, so that forests of little commercial value would be under the control of village *panchayats*. This led to a partial improvement in relations between villagers and the state, though in 1930 and 1931 there were many incidents of forest burning. Incendiarism usually occurred in areas designated by the government for the planting of *chir* pine.

The 1931 incidents coincided with nationwide interest in Gandhi's salt march to Dandi – a protest against British plans to tax the making of salt – leading to his arrest. There is no evidence that Gandhi himself expressed any opinion for or against the burning of forests in Kumaun, but some of the Congress leaders advised against it. Ten years later when, at the height of the Quit India movement, protesting villagers in Almora were killed by the police, Gandhi hailed their sacrifice.

The forest struggles in both Garhwal and Kumaun represented a conflict of interest between two opposing world views: commercial profitability and strategic imperial requirements on the one hand, and the traditional symbiotic relationship between village communities and natural resources on the other. In Tehri Garhwal the presence of the king gave a measure of legitimacy to attempts to repeal unjust laws, while in Kumaun there was a more rapid polarization between villagers and the colonial state. In neither region did the national movement play a significant role, at least until shortly before Independence, though the charismatic if distant activities of Mahatma Gandhi may have awakened a responsive echo.

Chipko and Appiko

In spite of the wide support which existed at the time of Independence for Gandhi's view of development, with its emphasis on *sarvodaya* and participative decentralized decision-making, the national leaders adopted an extremely different approach. Nehru believed that in order to be as strong as her former oppressors India must industrialize. This meant that raw materials such as water and timber would have to be available as cheaply and rapidly as possible.

Post-Independence forest policy reiterated the main tenets of the 1894 colonial Act, reinforcing the right of government to control all aspects of forest management. Princely states such as Tehri Garhwal had by then been abolished, which meant that the new government controlled an even larger domain than its colonial predecessor. The post-Independence era of the expansion of commercial forestry led to extensive roadbuilding throughout the Himalayas to facilitate the selective felling of commercial species. New uses such as paper manufacture were found for trees which were not previously marketable.

Selective felling did not produce sufficient yields of timber and it was therefore decided to embark upon larger-scale clear felling followed by the planting of quick-growing exotic species such as eucalyptus. Foresters were persuaded to grow teak and rosewood for the manufacture of furniture which could be exported to earn foreign exchange which, in a developing economy, is extremely valuable. Mixed tropical forests in the Western Ghats were clear felled and replaced by eucalyptus, tropical pine and teak. This phase of forestry was pursued between 1960 and the mid-1980s.

From 1975 onwards industrial forestry expanded onto farm lands and village commons under programmes of social forestry and wasteland development. These afforestation programmes were partly a response to the increased demand for timber, resin etc., but were also a reaction to public concern at the progressive denudation of hitherto forested areas. Social forestry involves the planting of trees on individual farms, the growth of trees for domestic fuel on government land, and community forests managed by village *panchayats*. Many farmers welcomed the introduction of eucalyptus as part of these programmes because of its commercial value and the fact that it is less labour intensive to grow. Where eucalyptus displaced food crops, however, the consequent rise in food prices plus the impact of increased unemployment as a result of the displacement of labour increased rural poverty.

The Chipko movement emerged from the struggles between villagers and the state over the latter's increasingly intrusive forest policies. A devastating flood in 1970 in the Alakanda valley, which stretches from Badrinath through Gopeshwar and leads eventually into the Ganges, was believed to demonstrate the connection between deforestation and floods. Protests were held in Gopeshwar the following year, and in 1973 the refusal of the forest department in Mandal to allocate ash trees to villagers while at the same time allowing a commercial firm to cut them, led to a call by Chandi Prasad Bhatt, a leading activist, to embrace (*chipko*) the trees to protect them. The loggers were obliged to turn back and were similarly unsuccessful at a second felling site.

In 1974 an attempt by contractors to fell trees at Reni was foiled by village women. A committee which included Bhatt was set up, and it was officially acknowledged that deforestation was responsible for the severe effects of the 1970 flood. Felling in the region was therefore halted for ten years. It was also agreed to monitor the tapping of resin to prevent permanent damage to trees. Later that year Sunderlal Bahuguna undertook a two-week fast at the Hanuman temple in Uttarkashi as a protest against the auctioning of forests, and young people from Garhwal and Kumaun went on a 700-kilometre march in support. An earlier *padayātrā* lasting a month had been undertaken by a swami from Rishikesh. Ghanshyam Raturi composed supportive poems in Hindi, one of which translates as follows:

Embrace the trees and
Save them from being felled;
The property of our hills,
Save them from being looted.

In Tehri Garhwal protests took place in 1974 against the excessive tapping of *chir* pine. A Sarvodaya leader undertook a fast, there were readings from the Bhagavadgītā and women tied *rakhi* bands around the wounded trees. (Hindu women traditionally tie threads around the wrists of their brothers once a year to signify the bond between them.) In Kumaun between 1974 and 1977 there were several confrontations between a newly formed activist branch of Chipko with Marxist leanings called the Uttarakhand Sangharsh Vahini and the foresters, who were supported by the police.

In 1978 the women of Pulna confiscated the implements of labourers about to fell trees; another group of women in the Pindar valley prevented their menfolk from selling an oak forest. Felling was also prevented at Parsari. Later in the same year Bahuguna led a campaign against logging in the Malgaddi forest. He went on a hunger strike and was interned in Tehri jail. More than 3,000 men, women and children took part in the campaign, which was accompanied by music and readings from the Gītā. The logging was called off.

In 1980 Bahuguna met with Prime Minister Indira Gandhi. He was able to persuade her to ban all commercial felling in the Himalayan forests of Uttar Pradesh for fifteen years. He began a long and arduous *padayātrā* along the full length of the Himalayas, from Kashmir to Kohima in Nagaland, to encourage people to understand and protect their forests.

The Appiko movement started in 1983 as a reaction against the felling of trees in Uttara Kannada. Young men and women members of the Lakshmi

Yuvak Mandali claimed that the number of trees marked for felling in a particular area was excessive and that in the course of felling the damage done to the trees and topsoil in dragging them was unnecessary. They decided to embrace (*appiko*) the trees to protect them.

The objectives of the Appiko movement were to protect existing forests, to regenerate trees in denuded areas and to give priority to conservation. To implement these aims local environmental centres were set up. These centres, known as Parisara Samrakshna Kendras, have monitored the activities of the forest department to check on infringements of the rules and have generally been able to raise environmental awareness in the Western Ghats.

In 1987 more than 100 voluntary organizations mainly from the southern states organized two simultaneous marches from both ends of the Western Ghats, to meet in Goa. The total length of this double *padayātrā* was 3,200 kilometres and the marchers visited more than 600 villages. Their intentions were to study the environmental problems of the Ghats and to mobilize villagers to prevent further destruction. Similar marches were subsequently organized in the Nilgiris, the Sivaliks in Jammu and Kashmir and in the Eastern Ghats of Orissa and Andhra Pradesh.

In 1989 a different kind of *padayātrā* was organized to converge on Kanyakumari in the extreme south to protest against the pollution of coastal waters and the destruction of river estuaries. One hundred women led the march. Their slogan was 'Protect waters, protect life'. The National Fisherfolk Forum summed up the aims of the *padayātrā* as follows:

1. to widen people's awareness of the link between water and life and to encourage popular initiatives to protect water;
2. to pressurize the government to evolve a sustainable water utilization policy and to democratize and strengthen the existing water management agencies;
3. to assess the damage already done, identify problem areas for detailed study and to evolve practices for rejuvenating water resources;
4. to revive and propagate traditional water conservation practices and regenerative fishing policies;
5. to form a network of all those concerned with these issues.

Women were at the forefront of all the activities that we have described so far. This is not surprising since it is customary in the Himalayan hill areas for women to work on equal terms with men in most agricultural activities except ploughing, in addition to being responsible for the collection of domestic fuel and fodder. However, their courage, as exemplified by the incident described at the beginning of this chapter and in the national struggle for independence, was exceptional.

I have drawn attention to several points of contact between the forest struggles up to and including the Chipko movement and Gandhi's national campaign. The *dhandak* method of protest was distinctive to the Tehri

Garhwal region, where the king could be called upon to redress injustices (Bahuguna's appeal to Indira Gandhi in 1980 is comparable to this approach). By the time of the 1921 agitations in Kumaun, Mahatma Gandhi was being invoked, perhaps somewhat unrealistically, to save the people from oppression, and the colonial state had become as evil as the legendary Rāvaṇa of the Rāmāyaṇa. (By this time Gandhi had started to describe British rule as satanic.) Ten years later Gandhi hailed villagers in Almora who were killed by the police as martyrs. By the 1940s there is no doubt that the self-confidence of the peasants of Tehri and Kumaun in their forest struggles was given a considerable boost by the success of the national movement.

During the post-Independence period the same kind of protests continued against equally repressive forest policies. There was contact with well-known Gandhians such as Mira Behn and members of the Uttarakhand Sarvodaya Mandal, which was founded in 1961 to promote forest rights, the organization of women, the abolition of alcohol consumption, and the establishment of local, forest-based small industries. We have also noted Bahuguna's Gandhian-style fasts and his lengthy Himalayan *padayātrā* – which may have been modelled on Gandhi's Dandi salt march – and the devotional use of the Bhagavadgītā.

Despite these associations between the forest struggles and Gandhi up to and including Chipko, Ramachandra Guha is reluctant to attribute too much Gandhian influence to them. He accepts that activists like Bahuguna and Bhatt 'exemplify the highest traditions of Gandhian constructive work', but feels that at the level of popular participation the Gandhian label is inappropriate because 'the role played by external ideologies is a severely limited one'. He maintains:

> Nor should superficial similarities in methods of protest lead one to designate Chipko as 'Gandhian', its 'non-violent' method being an inspired and highly original response to forest felling rather than ideologically motivated.[12]

However, it may also be argued that the true spirit of Gandhi is to be found not in 'isms' and ideologies, but in precisely the same kind of innovative, imaginative spontaneity that characterized his activities – in which case the spirit of the Mahatma and the spirit of Chipko are not so far apart.

Sunderlal Bahuguna

The emergence of the Marxist Uttarakhand Sangharsh Vahini during the Kumaun struggles of the mid-1970s is indicative of the range of viewpoints within the Chipko movement. In keeping with Marxist ideology, this wing believed that the economically powerful destroy nature for profit, and that both wealth and natural resources need to be redistributed. Marxists tend to be sceptical about tradition and optimistic about science and technology. A second and more socialist position is represented by Chandi Prasad Bhatt, who undertook afforestation programmes and pioneered appropriate technology.

Sunderlal Bahuguna represents a third major position, which is essentially Gandhian, within the Chipko umbrella. Bahuguna and his associates share with the Marxists a concern to remove social and economic injustices, and with Bhatt a more participative and less centralized view of society. In other respects the Bahuguna wing of Chipko is quite distinctive, however, and since it is currently the largest and most articulate section, we shall consider it in more detail in the person of its main representative.

Bahuguna's involvement in the Himalayan forest struggles began in 1973 when he was working in his home area as a social activist. We have already noted some of his views about ecological destruction in the Himalayas, which he attributes first and foremost to the legacy of colonialism:

> Most countries in south-east Asia have inherited a colonial system of administration and have adopted a development strategy imported from the western countries, whose economy was built on the basis of exploitation of nature and the colonies . . . Wanton exploitation of nature has created serious environmental problems like air pollution, water pollution and soil erosion. Development needs to be redefined.[13]

According to Bahuguna, the first principle of development is that those who wish to facilitate it must identify with common people:

> The first principle of field work is the identification of the worker with the masses. Mahatma Gandhi, the most successful social worker of the twentieth century, in his daily prayers said, 'Oh God, give me strength and eagerness to be one with the masses of India'.[14]

Surprisingly, in spite of his overall indebtedness to Gandhi, it is to the teaching of the Buddha that Bahuguna turns for his definition of development:

> The quest of social revolutionaries has been to make the life of the individual and society better. Gautam Buddha . . . was the first social revolutionary, who pondered deeply on this problem. He wanted to alleviate human miseries. His first discovery was that miseries cannot be alleviated from the palace – the illusion under which present day rulers keep the masses. So he left the palace and became a common man. He experienced all agonies including hunger, and reached the final stage of realization. Buddha realized that the root cause of misery was *tṛṣṇā* (desire). He defined development as a stage in the life of individuals and society in which they enjoy permanent peace, happiness and fulfilment.[15]

Freedom from desire is also expressed in terms of the doctrine of *niṣkāma* karma in the Bhagavadgītā.[16] We noted earlier the extent to which readings from the Gītā were used during Chipko activities. Bahuguna brings together the Gītā's three paths to liberation by comparing them with the work of scientists, activists and poets: 'The scientists represent knowledge (*gyan*), the activists action . . . and the literary men devotion (*bhakti*).'[17]

Bahuguna shares with Gandhi a mistrust of western notions of progress, but whereas Gandhi's attitude to science is ambivalent, Bahuguna believes in 'a science which accepts our role as caretakers of the Planet'.[18] Chipko exemplifies the best in both science and culture:

> The Chipko movement had its basis in Indian culture as well as in the scientific truth of life in trees, as proved by Dr J. C. Bose. This synthesis of science and culture in a holistic spirit can be extended to other spheres, specially in activities where the use of natural resources is involved.[19]

Bahuguna, like Gandhi, works in a highly-charged, symbolic manner, resorting to fasts and lengthy *padayātrās* to communicate his message. He is not an advocate of Gandhian ideology, but one who lives and acts for the moment: as Gandhi himself put it, 'I have no "ism", no ideology; my life is my message.'

Bahuguna believes there should be a total ban on felling and that forests should once again become the domain of village communities for use as fuel, fodder and fertilizer. Nationally and internationally, the industrial economy should be replaced by alternative modes of production in keeping with the principles of *sarvodaya*.

Like Gandhi, Bahuguna is able to address secular issues from within a spiritual framework shaped by Hindu classics such as the Bhagavadgītā and the inspiration of Vivekananda and the Buddha. His understanding of science has more in common with the other main branches of Chipko, represented by Bhatt and the Marxists, than with Gandhi's view of it, and he is familiar enough with the work of Indian scientists such as Jagadish Chandra Bose to recognize that science can take root as well in Indian as in western soil.

Anna Hazare

Kishan Baburao Hazare, better known as Anna ('elder brother') Hazare, is responsible for a remarkable programme of social and environmental uplift in the Ahmednagar district of Maharashtra. Although his work on village reconstruction has involved extensive tree planting, his most important initial contribution was to provide effective water management to one of the most drought-prone areas of the Deccan plateau.

Water management in ancient India was at an advanced state by the time of Kauṭilya's Arthaśāstra (see Chapter 2), which refers to irrigation canals and *bandhyas* (stoppages). Irrigation was an integral part of the political economy of the Mauryas, but the classic exposition of Hindu approaches to water management is the Bṛhat Saṁhitā of Varāhamihira, composed in the sixth century of the Common Era. This contains details of every aspect of the relationship between ground water and surface features, involving a wide range of physical and biological data.[20]

Anna Hazare was born in the region in which he now lives. The son of a poor labourer, he was able to obtain a reasonable level of education by moving to Bombay (Mumbai), where he enlisted for the army in the early 1960s. During the 1965 war with Pakistan he was the sole survivor in a border exchange of fire while driving a truck. The experience had a profound effect on him. Each year he took his annual leave in Ralegan Siddhi, where he was distressed by the progressive deterioration of village life and the increase of alcoholism. He left the army in 1975 in order to become a social worker.

Anna Hazare's water management schemes and his belief that 'rain water should be trapped where it falls' are in line with traditional Indian methods of raising the level of water tables. Under his guidance, six successful *nallah bandhs* (stream embankments) were initially constructed by a group of villagers in Ralegan Siddhi and a faulty percolation tank installed by the government was repaired.

From these small beginnings in 1975, the once poverty-stricken village of Ralegan Siddhi has been transformed into a model community which has become a source of inspiration and hope to many. The ground water table has risen considerably, 400,000 trees have been planted, an area which once produced a bare 30 per cent of its food requirements is now exporting food, and schools and hostels have been built and paid for.

During his army years Hazare had become acquainted with the work of Swami Vivekananda. Of one particular book he says:

> Swamiji writes that each individual strives for happiness and satisfaction. But each individual's definition of happiness is different and therefore each one tries to obtain it in a different way. The best way to obtain happiness is through service to others. I found the idea of service very appealing. Swami Vivekananda's books dramatically changed my attitudes to my own life and towards society.[21]

From Vivekananda, Hazare also derived his belief in the importance of sharing one's wealth with others, and of *sangha shakti*, the strength that is based on people working together. Villagers are encouraged to undertake work themselves in order to cut costs and to instil a sense of pride and collective ownership.

Hazare's attempt to break the vicious circle of unemployment, debt and alcoholism began with the reconstruction of the dilapidated village temple, which seemed to symbolize the condition of the village. Investing his army savings in the rebuilding, he called upon the villagers to supply voluntary labour for the task. Young people were particularly supportive of him, and he came to value what he describes as *yuva shakti*, the power of youth:

> God is everywhere but a child is first introduced to him in the temple. It is here that he receives education on the important values and morals of life. In a broader sense, the village itself is a temple where people serve and pray and learn the meaning of life.[22]

The temple became the centre for young people and Hazare took up residence there. Village disputes were often resolved there. Hazare used one particular dispute which had resulted in a drunken brawl as an opportunity to challenge the brewing of illicit liquor and alcoholism. He found alternative employment for most of the liquor brewers and had the few remaining brewing units smashed by the village youth. Persistent drinkers were tied to trees and flogged. One offender was left tied to a pole for two days. Asked what Gandhi would have said of such rough justice, Hazare replied: 'When Gandhiji's words prove inadequate, it becomes necessary to use Shivaji's sword to fight the evils.'

The irrigation schemes which benefited the 70 per cent of the villagers who owned land were made possible through bank loans taken out by village cooperatives. Some villagers had great difficulty in repaying their contribution. Fuel and fodder were provided through afforestation via the social forestry scheme of the Department of Social Forestry. Children nurtured the saplings. Cattle were provided with grazing, but unproductive cows were sold for slaughter.

Community toilets have been introduced and used in conjunction with biogas plants which supply methane for domestic cooking. Smokeless *chulas* (stoves) diminish the need for firewood and have reduced the incidence of respiratory illnesses. The school hostel has a solar water heater and there are solar street lights which were installed mainly for demonstration purposes. A windmill pumps water from a well. Women play a large part in all the village development activities and a women's *panchayat* has been set up.

During the last few years Hazare has taken up cudgels against corrupt practices in the state government. He undertook a fast in protest against irregularities in the Department of Social Forestry, and in 1991, on the anniversary of Gandhi's birthday, threatened to return a national award for his work to the President of India if his demands were not met. More recently he has taken on the 'saffron brigade' (Shiv Sena and BJP) in Maharashtra. His campaigns are widely reported.

> Anna is agitating . . . against corruption. Only, this time round the avengers of 1995 – the saffron brigade – are the accused. The Sena-BJP alliance has been making desperate efforts to win over Hazare again.[23]

In September 1998 he was jailed for three months on a charge of criminal defamation for accusing the state social welfare minister of corruption.

Such agitations against corruption in high places are reminiscent of Gandhi's tactics against colonial rule. In other respects, however, Anna Hazare appears less Gandhian than, say, Sunderlal Bahuguna, who challenged the entire industrial-commercial onslaught against the forests. Hazare has more in common with Chandi Prasad Bhatt, who has proposed socio-technological alternatives to environmentally degrading technologies. If Bahuguna challenged the forces of oppression so that Bhatt could demonstrate in practical terms what might replace them, Anna Hazare worked the other way round. He began by demonstrating what could be achieved by a combination of human energies and technical know-how – in part traditional. He then tried to reform the corrupt superstructure which inhibits the very processes that it is designed to facilitate. Before he attempted any of these things, however, he rebuilt the village temple.

We have considered the manner in which organized resistance to the commercial destruction of forests in the lower Himalayas grew out of attempts to redress grievances by traditional non-violent methods such as the *dhandak*. Such peasant revolts were protests against the loss of their source of livelihood and of their entire existence as a community. Not only were vast areas of mixed forests of broad-leaved trees such as oak destroyed, but their replacement by commercially valuable species such as *chir* pine further disturbed the ecological balance, leading to floods, soil erosion and the silting of rivers.

The Chipko and Appiko movements grew out of these earlier protests and reflect the complete incompatibility between the commercial and industrial interests of the state and the needs of ordinary people, who have traditionally enjoyed a symbiotic relationship with their natural surroundings. They are not, therefore, environmental movements in the western sense (e.g. when protesters oppose the expansion of an airport), but a total response to the destruction of communities and their habitats. Within these movements we have seen the powerful use of religious symbols such as the *padayātrā*, fasts and devotional readings from the Bhagavadgītā, and the manner in which the various leaders emulated Gandhi by addressing secular issues from within a spiritual framework. We have also noted the influence of Jagadish Chandra Bose on Sunderlal Bahuguna and of Vivekananda on Anna Hazare.

Within the Chipko movement three kinds of ideology have emerged. There are the predominantly young Marxists who believe that environmental degradation is due to the profit motive of the rich plus unequal access to resources. They organize the poor to work for the redistribution of power and see modern science as a means to a better quality of life for all. An economically just society must be the basis for sound social and ecological relationships.

Gandhians, represented by Sunderlal Bahuguna, are committed to the *sarvodaya* ideal of participative rural communities, and view the natural world with traditional Hindu reverence. They reject socialism as western and are critical of what they see as 'the fallacy of increasing labour productivity independently of the social and material context'.[24]

The Gandhian view of the relationship between resources and needs may be summarized by Bahuguna's claim that 'ecology is permanent economy'. Most modern Gandhians, including Bahuguna but especially representatives of C. P. Bhatt's type of ideology, are more optimistic than Gandhi about the potential of science for improvement. However, the science that is most appropriate for eco-restoration need not necessarily be either modern or western, as Anna Hazare's programme of water management makes clear.

Many of the Hindu elements in India's nascent environmental movement represent continuity with the past. We may cite the use of readings from the Bhagavadgītā, the *dhandak*, the characterization of centralized bureaucracy as the legendary Rāvaṇa, and Anna Hazare's water harvesting techniques as examples, but there are also some imaginative transformations. These include the embracing of trees to indicate that they, like people, are living, and the related practice of tying *rakhi* bands around trees to represent a bond of kinship with them.

Other examples of the transformation of tradition include Chandi Prasad Bhatt's advocacy of appropriate technology which, though less sophisticated than J. C. Bose's work, represents a move in the direction of environmentally sensitive public interest science (as opposed to 'scientific' forestry). Sunderlal Bahuguna's *padayātrā*s are an imaginative reinterpretation of the traditional notion of pilgrimage, infusing it with an added educational and political dimension.[25] Anna Hazare's use of his village temple as the focus for eco-development is an example of the way in which a traditional symbol of Hindu orthodoxy can be transformed to enable it to play a vital contemporary and ecumenical role.

5 Ecology and Buddhism

In 1960 the journey from the Indian border to Thimpu, Bhutan's capital, took five or six days by mule. Today it takes about six hours by car along a 200-kilometre mountain road. From the deciduous woodlands of the humid, subtropical south the road rises through the mixed, broad-leaved forests of central Bhutan up to the coniferous evergreens of the north. In the valleys where there is less moisture, grass and scrub bushes predominate. Beyond Thimpu lie the Great Himalayas, where alpine shrubs and rhododendrons give way to bare rocks and icy tundra. The entire country is a little larger than Switzerland.

A senior minister of Bhutan's National Environment Commission recently summed up the country's environmental priorities as follows:

> Bhutan was concerned about the environment long before the Earth Summit. In the sixties our forest situation was beginning to get out of hand, but in the seventies we took stock of it. Our dams use run-of-the-river schemes in which there is no loss of habitat or adverse effect on aquatic life. The electricity they produce is sold to India and brings us a large revenue.
>
> Wind power is being proposed and we are working on solar power with Dutch help. We have a waste project which will send recyclable waste to India. We monitor air quality and have banned reconditioned vehicles. Laws have been introduced to control exhausts.
>
> In 1986 we had a major programme of environmental awareness raising in schools. The King was very supportive of this. Monks can play an important part in raising environmental awareness – the community listens to them. They are already being used by the health department to teach people.[1]

At first sight the combination of technology and Mahāyāna *tantra* – the official religion of Bhutan – as a means of raising social and environmental awareness may seem unusual. However, Buddhist *tantra* is essentially very practical and this-worldly, and there is no doubt that the Bhutanese monks who are involved in a wide variety of secular activities are very committed to every aspect of their country's welfare. According to Bhutanese statesman and scholar, His Excellency Om Pradhan:

> The Buddhist faith permeates all strands of secular life, bringing with it reverence for the land and its wellbeing. Annual festivals . . . are spiritual occasions in each *dzongkhag* [province]. They bring together the population . . . and are dedicated to either Guru Rimpoche or other deities. Throughout Bhutan *stūpas* and *chortens* line the roadside.[2]

I shall say more about Bhutan in a later section. In the meantime we must explore the main characteristics of Buddhist teaching and its historical development in order to evaluate those features of contemporary Buddhism which represent continuities with or transformations of the past.

Early Buddhism

By the time of the Buddha's birth, which recent scholarship suggests to be around 480 BCE, the focus of Aryan domination had moved from the Punjab to the Gangetic valley. Sections of the forested areas of what are now Uttar Pradesh and Bihar had been 'eaten up' by Agni to provide for agrarian settlements, which probably consisted of single units or small amalgams of tribes surrounded by tracts of wild and dangerous terrain. As agriculture improved, food surpluses enabled the Aryan élites to overrun the tribal gatherers to whom they assigned an inferior status.[3] In time these surpluses promoted trade to such an extent that powerful states such as that of the northern Mauryas came into being.

Siddhārtha, who became the Buddha ('enlightened one'), was born in a sacred grove of *sal* trees in Kapilavastu, in what is now Nepal. A *kṣatriya* from an aristocratic family, he abandoned home at an early age in search of enlightenment, which he achieved in an animal park in Bihar. After many years of travelling and teaching, he died at the age of 80.

So much is probably fairly accurate history. The reconstruction of early Buddhist teaching is more difficult. The main theme of the Buddha's first sermon following his enlightenment is the Four Noble Truths. The first of these is the recognition of the universal fact of *dukkha*, which is the transience or unsatisfactoriness associated with such things as old age, sickness and death (in which other sentient beings also share). *Dukkha* is part and parcel of our personhood, which is described in terms of five *khandhas* or aggregations. These are form (the physical aspect of the world), sensing (the ability to sense feelings), perceptions (the labelling of sense objects), impulses (reactions to perceptions – conscious or otherwise) and consciousness, denoted by *viññāna* – a term we shall encounter later. These five *khandhas* are constantly changing to such an extent that we cannot speak of a self, in the Hindu sense. Hence the Buddhist doctrine of no-self, denoted by *anattā* (which the Mahāyāna further develops in terms of 'emptiness').

The second Noble Truth recognises that *dukkha* is caused by craving for satisfaction from things that are impermanent (*anicca* – another important concept). The third Truth acknowledges that *dukkha* can be brought to an end by the removal of craving, and the fourth sets out the way to do this, which is the Noble Eightfold Path. The ending of *dukkha* is denoted by *nibbāna*.

The Noble Eightfold Path is not a sequence, and all eight components should ideally work together. Correct insight and resolve mean that a person is familiar with Buddhist teaching and wishes to proceed further. Appropriate speech, conduct and livelihood are moral and social necessities, and correct effort, mindfulness and meditation lead to the direct perception of the nature of reality and of the chain of causation responsible for our present existence. Buddhist causality is denoted by *paṭicca samuppāda*, which we translate as 'interdependent co-arising', a cardinal doctrine with important ecological connotations.

The Four Noble Truths and the Noble Eightfold Path are the central teaching of early Buddhism. The fivefold principles known as *pañca-śīla* prohibit harm to any living being, stealing, sexual misdemeanour, lies and insults and the taking of intoxicants. They apply to all Buddhists. A novice must observe an additional five precepts (plus a total ban on sexual activity), a lay nun must observe eight (e.g. in Thailand), and a monk must fulfil all the 227 rules of monastic discipline, known as the Pāṭimokkha.

Monasteries began as a convenient alternative to homeless mendicancy during the rainy season, and the first scriptures were probably codified in them. The Vinaya (which contains the Pāṭimokkha) and Sutta constitute the earliest group of scriptures in the Pali canon. The Mahāyāna, however, adds more, some of which have become foundational for particular schools of Buddhism. Most of these were composed in Sanskrit, but have survived only in Chinese and Tibetan.

Accounts of the Buddha's life are richly embellished with allusions to nature. As he took his first steps, lotus flowers sprang up. During childhood he often meditated beneath a jambo tree – a species of myrtle. His enlightenment took place under the spreading branches of the *bo* tree – sacred to Buddhists, Hindus and Jains. When he departed this life, *sal* trees blossomed out of season.

In an early text, the Buddha says:

> Know ye the grasses and the trees . . . Then know ye the worms, and the moths, and the different sorts of ants . . . Know ye also the four-footed animals small and great, the serpents, the fish . . . the birds . . . Know ye the marks that constitute species are theirs, and their species are manifold.[4]

The earliest monastic rules enshrined in the Pāṭimokkha contain numerous injunctions against environmental irresponsibility. Some are basically sound advice governing personal and communal hygiene, but others are designed to avoid harm to sentient beings. Thus a monk may not cut down a tree or dig the earth (because that would destroy small life forms) and he must not empty a vessel of water containing, say, fish, onto the ground.[5]

The Jātaka accounts of the Buddha's former births give a good idea of the extent to which Buddhist teaching urges respect for animals. Christopher Chapple has drawn up a list of all the animals that are to be found in these narratives. Of the 550 stories accepted as canonical by Theravādins, half mention animals as the central characters. Seventy different types of animal appear with varying frequency: monkeys are present in twenty-seven tales, lions appear nineteen times, and the humble mouse just once![6]

The Buddha rejected the Hindu notion of self (*ātman*), though not the existence of non-embodied spirits, and many Buddhist scriptures contain references to spirits residing in natural objects. The Aṅguttara Nikāya tells the story of a man who sheltered under a *banyan* tree and ate its fruit, then broke off a branch and went away. The story continues:

> Thought the spirit dwelling in that tree: how amazing, how astonishing it is, that a man should be so evil as to break a branch off the tree after eating his fill. Suppose the tree were to bear no more fruit. And the tree bore no more fruit.[7]

The activities of the Maurya state under Candragupta are recorded in the Arthaśāstra, which we noted in Chapter 2 for its detailed preoccupation with the conservation of natural resources, including forests. Not only did the state control cultivated land through peasant intermediaries, but it began to annex hunting lands which had been the traditional preserve of food-gathering tribals. The Mauryas conserved resources even more by promoting the protection of rare plants and animals, sacred groves, ponds and rivers.

Following his conversion to Buddhism, the Maurya emperor, Ashoka, carried out large-scale plant and forestry improvements which are mentioned in his pillar edicts. One of these, published in the third century BCE, in what is now Orissa, reads as follows:

> The king . . . enjoins that: medical attendance shall be made available to both man and animal; the medicinal herbs, the fruit trees, the roots and tubers, are to be transplanted in those places where they are not presently available . . . wells should be dug and shadowy trees should be planted by the roadside for enjoyment both by man and animal.[8]

Another of Ashoka's pillar edicts effectively marks the end of the Aryan method of land-clearing: 'Forests must not be burned either uselessly or in order to destroy [living beings].'[9]

Whereas Kauṭilya's Arthaśāstra had set out a view of the state whereby the king performs his royal *dharma* in order to work out his personal salvation, Ashoka is recorded as having assumed a more cosmic role:

> When kings are righteous ... moon and sun go right in their courses ... seasons and years go on their courses ... and the sky-*deva* bestows sufficient rain. Rains fall seasonably, the crops ripen in due season.[10]

The differences between Kauṭilya and Ashoka should not be overemphasized or attributed too readily to Hindu (brahmanical) and Buddhist influence, since both were based on a common stock of Indo-Aryan politico-moral ideals. The tendency during the Ashokan period was to view the king as the vehicle for the flow of the eternal, universal *dharma*, giving his rule a unique quality. Early Buddhist scriptures draw parallels between the Buddha, who was believed to turn the *cakra* (wheel) of *dharma*, and the king, sometimes a *cakravartin*, who bore the wheel insignia. Ashoka and his Buddhist successors, mostly in other parts of Asia, became known as the turners of the wheel of cosmic law. (Some scholars doubt that Ashoka ever claimed to be a *cakravartin*.)

Buddhism reached Sri Lanka during the reign of Ashoka largely via his son, Mahinda, who had become a monk. There the Ashokan pattern of king, monastic community (*saṅgha*) and people was maintained and eventually exported to southeast Asia, but in some cases this did not occur until Mahāyāna Buddhism had already arrived there. The Theravāda and Mahāyāna traditions may well have developed in parallel from an earlier common core. The former preserves the tradition of the elders (*theras*), whereas the latter reinterpreted early Buddhism in a variety of imaginative ways. The compilation of what became the Pali canon (the Theravāda text collection) started in India and was probably completed in Sri Lanka by the end of the first century BCE.

The Mahāyāna

Mahāyāna Buddhism regards itself as a more complete expression of the *dharma* than the 'lesser' Theravāda. Its distinctive teaching is centred on compassion for all sentient beings to such an extent that the ideal Mahāyānist will delay his or her entry into *nirvāṇa*-beyond-rebirth until they are all able to do the same. A key concept, which played a major role in the development of Mahāyāna Buddhism, is *śūnyatā* – emptiness or voidness. Just as the no-soul doctrine denies the existence of an enduring soul, so *śūnyatā* stipulates that the whole of reality is empty. According to Nāgārjuna (150–250 CE), emptiness defines the true character of all reality:

> To the untrained mind, reality appears to consist of subjects, objects, and their relationships, all existing in their own right ... which are therefore clung to or avoided as the real sources of happiness or misery. When

these entities are seen as space-like rather than particle-like, clinging and avoidance cease and liberation results.[11]

Nāgārjuna's idea of *śūnyatā* was that things are empty of inherent nature or essence, since their existence and nature are radically dependent on other things. He recognized that although conventional language speaks of appearances according to their characteristics, in fact all appearances, since they are *śūnyatā* (empty of self), have the same nature. It follows that *nirvāṇa* and *saṁsāra* are the same; they cannot be other than each other because everything is empty of any distinctive nature or essence that might distinguish it. The purpose of life is not to strive for moral perfection, but to uncover one's true nature.

In the earliest stages of Buddhism the ideal Buddhist was known as an *arhat*, essentially one who practised insight into various attributes such as loving-kindness. The Mahāyāna replaced this notion with the *bodhisattva*, who was characterized by transcendental wisdom (i.e. wisdom that could see reality from the perspective of emptiness) and compassion for all living beings. The term *bodhisattva* is pre-Mahāyānist and popular early Buddhist literature contains many references to the concept. Transcendental wisdom is *prajñā-pāramitā*, which expresses a level of compassion so strong that some *bodhisattvas* refuse to enter *nirvāṇa*-beyond-rebirth until all sentient beings can do the same.

Where wisdom had been taught as supreme at an early stage of Buddhism, and compassion as secondary, the Mahāyāna ranked them equally. The apparent contradiction between undifferentiated emptiness and compassion for living beings was resolved by incorporating and transcending the contradiction. From the perspective of transcendental wisdom the *bodhisattva* sees no persons; out of compassion he (or she) resolves to save them. (*Bodhisattvas* can be feminine; wisdom is also feminine.)

A person becomes a *bodhisattva* when they resolve to win enlightenment for all creatures. After performing six perfections he or she can understand the meaning of emptiness and apprehend the nature of true reality (stage seven). Renouncing *nirvāṇa*-beyond-rebirth for the sake of others, the celestial *bodhisattva* cannot be prevented from gaining full Buddhahood (the step beyond stage ten). *Avalokiteśvara* is a *bodhisattva* at the ninth stage; repetition of the original vow at this point can generate considerable benefits for others:

> All creatures are in pain, [the *bodhisattva*] resolves . . . All that mass of pain and evil karma I take in my own body . . . I take upon myself the burden of sorrow; I resolve to do so; I endure it all . . . Assuredly I must bear the burden of all beings . . . I work to establish a kingdom of perfect wisdom for all beings.[12]

An early sect known as the Mahāsāṅghikas had diminished the role of the historical Buddha and elevated the long-enlightened Buddha, whom they conceived as perfect, infinite and permanently withdrawn into trance. Some Mahāyāna schools took over this Buddhology and attempted to restate it as the doctrine of Three Bodies (*trikāya*). Reality is viewed here as three aspects of the nature of Buddhahood corresponding to the realm of phenomena (*nirmāṇa-kāya*), with Buddhas located in history, the realm of highest truth (*dharma-kāya*), which is ultimate Buddhahood, characterized by emptiness, and the intermediate realm of bliss (*sambhoga-kāya*), filled with journeying *bodhisattvas* and other beings (gods, spirits etc.).

Nāgārjuna's views were carried forward by the Mādhyamikas, the influential second-century school which he founded and which claimed that *śūnyatā* is in between (*madhyama*) conflicting positions. They emphasized the idea of instantaneous continuity whereby phenomena constantly reproduce themselves in a moment-to-moment sequence of change. Thus the true nature of phenomena (of which we can only see the apparent continuity, as in a film) is *śūnyatā*. Since all 'emptinesses' are the same, then this particular emptiness must be equivalent to interdependent co-arising. Mādhyamika philosophy played an important part in the development of Chinese Buddhism.

The idealistic Vijñānavādins, also known as Yogācārins, were influential in the fourth century CE. Their main thesis was that consciousness is the only reality, and that the diversity of empirical phenomena derives from mental projections which are misinterpreted. The foundation of our personal identity is our storehouse consciousness (*alāya-vijñāna*), which has become tainted by bad karma, causing the illusion of the duality of an objective world separate from an observing subject. Through *yoga* we can come to realize that *alāya* is the only reality. This school paved the way for the *tantric* synthesis of the Vajrayāna, and played a large part in the development of Tibetan and Far Eastern Buddhism.

The spread of Buddhism

Buddhism arrived in Sri Lanka during the reign of Ashoka. The Sri Lankan king was reconsecrated according to the instructions of Ashoka, four monks having first been despatched to establish the *dharma* on the Island. Trevor Ling describes the Ashokan Buddhist state in India and that replicated in Sri Lanka as the first sustained realization of the Buddhist ideal:

> The *Saṅgha*, the new community of those who have abandoned the individualistic notions which nourish so much 'commonsense' understanding of life, and which produce so much envy, hatred, sorrow and conflict, constitutes the growing point – or growing points – of the restructured humanity. Meanwhile, the large remaining area of society outside the *Saṅgha* . . . must have its own appropriate forms of organization and control, which will both discourage the violent and morally unwholesome elements, and encourage the pursuit of peace and morally wholesome action. In ancient India this task had to be performed

by a Buddhist king, and this is the task that Ashoka appears to have accepted and endeavoured to fulfil, with notable success.[13]

In Sri Lanka the cult of *stūpas* (mounds which often contained sacred relics), *vihāras*, which housed monks and functioned as shrines (often combined with a *dāgaba* or relic chamber – not necessarily the same as a *stūpa*), sacred trees and *devālayas* (temples of Hindu gods), flourished as part of orthodox Buddhism. The famous Tooth Relic, claimed to be a left tooth of the Buddha and now housed at Kandy, became associated with the royal court.

By the end of the third century BCE all the essential components of the Ashokan state were present in Sri Lanka. There was a king who was a practising Buddhist, an indigenous *saṅgha* and a variety of *stūpas* and *bodhi*-trees to satisfy popular devotion. This pattern remained essentially intact until the political annexation of the island by the British in 1815. Some kings devoted their energies to the building and maintenance of shrines and support for the *saṅgha*, while others did more for the material well-being of their people. Dhatusena, for example, built a large reservoir from which dry areas more than seventy kilometres away could be irrigated, while Parakkama Bahu (twelfth century CE) devised a scheme to irrigate the whole island. Hospitals, homes for invalids, the blind and destitute women were set up and provision was also made for the care of sick animals.[14]

At the time Buddhism began to make a significant impact on China during the first and second centuries CE, Confucianism was in decline. Its fortunes had waxed and waned since its inception during the sixth century BCE and it was increasingly seen by the peasantry as an oppressive weapon in the hands of the rich. By contrast Taoism and Buddhism were open to all, and the assistance given by Taoist scholars to Buddhist monks greatly facilitated the assimilation of the latter's teaching. When the Han dynasty collapsed in 220 CE the non-Chinese invaders who overran the north were more willing to embrace Buddhism than either their Confucian or Taoist predecessors.

Both the major schools of Buddhism may have reached China at about the same time. Early texts such as the Vinaya transmitted the basic pattern of monastic discipline, and by the middle of the third century CE Chinese scholars such as Chu Shih-hsing were well versed in the Mahāyāna Wisdom literature. The task of translating from the phonetic system of the Indian languages into the ideographic Chinese script was enormous and would not have been accomplished but for the energy and industry of Kumarajiva (344–413 CE), who concentrated his efforts on expositions of the philosophy of Nāgārjuna.[15] The Chinese Tripiṭaka or Buddhist canon contains 1,440 works amounting to 5,586 volumes.

Although the Buddha was born in Kapilavastu in Nepal, it was not until Ashoka erected several *stūpas* in 249 BCE that organized Buddhism began to take root there. Theravāda monasteries and Śaiva temples coexisted until *tantra* made its way to Nepal from the Vajrayāna centres in Bihar. The Nepalese Buddhist canon includes many of the original works of the Sanskrit scriptures such as the Wisdom literature. As Buddhism declined in India, Nepalese Buddhism closed in on itself; caste distinctions began to creep in and monastic life became less rigorous. In the eighteenth century the Gurkhas, who trace their ancestry to the north Indian *rajputs*, conquered Nepal, and the state religion became Hindu.

It was not until the reign of the Tibetan king Srong-san-gan-po (seventh century CE) that spoken Tibetan was first accorded an alphabet, which was based on an Indian prototype brought to Tibet by Buddhist monks. However, Bön religion (which probably preceded Buddhism), with its emphasis on nature worship and evil spirits which can be controlled by spells and magic, was not compatible with either conventional monasticism or the more philosophically based Mahāyāna schools. In the eighth century, however, in response to an invitation from another Tibetan king, a teacher called Padmasambhava set out from India to establish what has become the most important form of Tibetan Buddhism, usually known as Lamaism (though Tibetans dislike the term). He did this by combining elements of *tantra* with a modified form of indigenous Bön beliefs. Padmasambhava (Guru Rimpoche) was the founder of the Red Cap Order (Nyingmapa) of Lamaism. The Yellow Cap Order (Gelugpa), which ultimately became more popular, had been founded by the eleventh century. It lays strong emphasis on monastic discipline and philosophical reflection, and may be seen as a reform movement within Tibetan Buddhism.

The iconography of Tibetan Buddhism is enormously complicated. The Yellow Cap Order, for example, acknowledges more than 300 divinities; of these the *bodhisattva* Avalokiteśvara has 108 forms. The Dalai Lama is seen as the living incarnation of Avalokiteśvara. The Tibetan Buddhist canon, which was more or less complete by the thirteenth century, runs to more than 300 volumes.

Southeast Asia

The first Buddhist emissaries to southeast Asia were sent by Ashoka in the third century BCE. They followed familiar trade routes to the 'Land of Gold', which was probably the west coast of Indonesia. By the beginning of the first century CE many Indians had settled in this region.

Mahāyāna Buddhism began to arrive in Indonesia in the seventh century CE, and a mixture of Sarvāstivādin and Mahāyāna Buddhism became influential in Sumatra under the patronage of the Srivijaya kings. Indonesian students travelled to Nalanda in Bihar where they became acquainted with Vajrayāna teaching. The magnificent Borobudur centre in Java combines the early Buddhist *stūpa* with a more exotic style of Vajrayāna architecture. *Tantric* Buddhism, together with brahmanical and Śaivite Hindu beliefs, coexisted in the region until the arrival of Islam in the fourteenth and fifteenth centuries.

The Tibeto-Burmese who have inhabited Burma since the eleventh century were preceded by the Pyu, who were of Tibetan stock, and the Mon, who are thought to have come from the east coast of India. The Pyu of central Burma had adopted Theravāda Buddhism – possibly exported from south India – by the sixth century, and the Mon, further south, were also Theravādin. Mahāyāna Buddhism may have arrived overland from Bengal and Assam even earlier.

During the eleventh century the king of Pagan attempted to remove *tantric* influences from Burmese monasteries and restore Buddhist monasticism to its original Theravādin form as preserved by the Mon. Mahāyāna Buddhism declined, and Pali superseded Sanskrit as the official language of the scriptures, which were revised with the assistance of monks from Sri Lanka. The unification of Burma as a great centre of Buddhist culture and learning during this period resulted in the suppression of many cults of *nats* or indigenous spirits, though both these and certain elements of brahmanism remain a significant feature of Burmese Buddhism.

The Khmer, who came from the Mekong Valley, assumed control of what later became Siam and Cambodia during the ninth century. They were Hindus who readily combined the worship of Vishnu and Śiva with Mahāyāna *bodhisattvas* and indigenous spirits and ancestors. Their king was believed to be the human representative of Śiva who reigned from Mount Meru which, according to Indian mythology, was the centre of the universe. Angkor Wat was built during the reign of Suryavarman II in the twelfth century. It represented both the climax and the end of the great classical and mainly Hindu art of the Khmer period. Khmer cosmology played a central part in the belief systems of both Buddhists and Hindus in southeast Asia.

The Khmer kingdom was overthrown in the fifteenth century by the Thai, who were Theravāda Buddhists living at that time in the vicinity of the Chao Phaya river. Under their influence Theravāda Buddhism became the dominant religion in Cambodia, though much of the earlier Hindu mythology was retained. Theravāda Buddhism was also introduced into Laos when it was founded in the fourteenth century. The Laotians are similar to the Thai and their culture exhibits a variety of popular elements such as astrology, which Buddhism had no difficulty in retaining.

The Thai, like the Burmese, have been less influenced by Hindu beliefs than Cambodians and Indonesians. This is primarily because the Mon peoples embraced Theravāda Buddhism comparatively early. The oldest and largest *stūpa* in Thailand is at Nakhon Pathom and dates from the sixth century. Between the eighth and the thirteenth centuries the region was dominated by Indonesian Mahāyānists, and from the eleventh until the fourteenth centuries the Hindu beliefs of the Khmer were influential. However, from the fourteenth century onwards, during which Thai and Sinhalese monks were in frequent contact, Theravāda Buddhism flourished.

From this brief survey it will be clear that the religious influences which shaped the individual countries of southeast Asia have been enormously complex and varied. However, if we omit Malaysia and Indonesia on the grounds that they are currently dominated by Islam (which lies outside our scope), certain features exist in common among the remaining countries.

In each of the four countries, Burma, Thailand, Cambodia and Laos, Mahāyāna Buddhism long ago diminished in influence, and whatever *tantric* elements may have been present at an earlier stage have been largely suppressed (e.g. in Burma). Brahmanism and cults centred upon the worship of Vishnu and Śiva have exercised a steady influence on the entire region, and especially in Cambodia, though these have been overshadowed even there by the growth of Theravāda Buddhism. Animism and ancestor and spirit worship have continued in many areas.

Theravāda Buddhism has grown in all four countries to such an extent that monks can move between monasteries without hindrance (unless for political reasons), using the same Pali scriptures and adhering to the same monastic discipline. Over the centuries they have established links with Sri Lanka and many of them continue to study at the centres of Buddhist learning in India. Each of the four countries has a chief monk who supervises the national monastic order.

The history of these southeast Asian countries has also been shaped by a cosmology and a concept of monarchy that incorporates Hindu and Buddhist elements, ranging from the god-king (*devarāja*) and *bodhisattva* of Cambodia to the Ashokan *cakravartin*, which is common to most Theravādin countries. It is this aspect of Theravāda Buddhism and the accompanying relationship between king, *saṅgha* and society as a whole that provide the ecumenical framework for contemporary expressions of social and environmental concern. We shall consider this in more detail in the next chapter, returning for the time being to north India.

Ladakh

In this and the next section we shall consider the two Himalayan regions of Ladakh and Bhutan in relation to connections between Mahāyāna Buddhism and ecology. In each case the argument is that Buddhism, by virtue of its naturalistic and this-worldly outlook, enshrined in a variety of activities based on monasteries in which monks, nuns, and their lay adherents share a keen awareness of the importance of living in harmony with nature, provides a fruitful context for dealing with social and environmental problems in a most ecologically important part of the world.

Though administratively part of Kashmir, Ladakh ('the land of mountain passes') was an independent kingdom until 1834, when the Hindu Dogras invaded it and it came under the jurisdiction of the Maharajah of Jammu and Kashmir. The advent of the Chinese into Tibet in the 1950s and across the border into India in 1962 made Ladakh an important military zone.

Ladakh is a dry plateau covering an area of 100,000 square kilometres, which in 1990 contained a population of about 130,000. Many ethnic groups in Ladakh have combined over the centuries to produce an overall culture which is Tibetan, but which also includes Shi-ite Muslims in the west and

Sunnite Muslims in the capital city of Leh. Overall, the western district of Kargil is largely Muslim and the district of Leh is mainly Buddhist, with a few hundred Christians in the city itself. Buddhism may have reached Ladakh as early as the second century BCE. All branches of Tibetan Buddhism are currently represented, and the Dalai Lama is acknowledged as spiritual leader.

Ladakh's magnificent monasteries form the country's backbone of culture and social organization. Many of them own a substantial amount of land, which farmers cultivate in return for produce. They can also serve as grain banks. Young men who cannot be supported by their families are ordained as monks and women can similarly become nuns. Monks perform religious ceremonies in individual homes as well as in the monasteries, and are increasingly paid for their services in cash. The social importance of Ladakh's monasteries makes them ideal for the demonstration of environmentally appropriate technologies: the Matho monastery possesses a hydraulic ram pump which lifts water 50 metres to the top of the hill on which it stands.

The essential teachings of Buddhism, usually in its Vajrayāna form, are contained in the art and sculpture of the monasteries and in the seasonal dramas and dances. The colourful *cham* dance, for example, culminates in the ritual destruction of egocentricity and selfishness that are the enemies of spiritual liberation. Vajrayāna teaching emphasizes enhanced psychic experience here and now, rather than the distant realisation of *nirvāṇa*, and few Ladakhi monks seem to pay serious attention to any life beyond this one. The prevalence of Bön beliefs means that there is also considerable interest in spirits and the occult.

Many religious occasions are celebrated in people's homes. On the tenth day of every month, for example, villagers gather in their houses to honour the birthday of Guru Rimpoche, who first introduced Buddhism into Tibet. They read religious texts which often include the Wisdom literature (Perfect Wisdom or *prajñā-pāramitā*), with its emphasis on compassion for all living beings and the *bodhisattva* who refuses to enter *nirvāṇa* until all beings can be saved. According to Helena Norberg-Hodge, who has written extensively about Buddhism in Ladakh, they are familiar with the notion of *śūnyatā*, which she interprets as the view that all things ultimately dissolve into a web of relationships.[16]

Many Ladakhi houses display prayer flags whose colours correspond to wisdom, strength and compassion – the attributes of Perfect Wisdom – and there are artefacts representing a variety of spirits. Red dots and swastika-like symbols are designed to placate the *tsan* spirit who rides a white horse. Fair to behold face to face, he can inflict serious injuries when his hideous back is turned.

The Mahabodhi International Meditation Centre is a few kilometres outside Leh and was established in 1986. In common with other Mahabodhi centres, which have grown out of the efforts of Sri Lankan Theravādin monks to restore Buddhism in India, this centre combines a wide range of spiritual, social and medical activities. When I visited the centre in October 1995, monks in residence at the centre included Yellow Cap Gelugpas and Red Cap Nyingmapas, Karguyrudpas, Drukpas, Digungpas and Saskyas (the centre itself is non-sectarian).

The nuns at the Mahabodhi centre were of two types: *chomu*, who live and work at home, and *gyetsulma*, who live at home but undertake more specialized activities such as teaching. *Chomu* can be fully ordained only in Taiwan after a period of training there. A representative of the centre explained that there were plans to build a nunnery to accommodate twenty-five nuns and provide a good religious and secular education for them in a monastic environment. Their main task would be to support and encourage backward Ladakhi women. At the time of my visit there were estimated to be about 400 nuns in Ladakh.

The centre maintains a residential school for poor children, a mobile health clinic, a home for the aged and a charity programme, and there are plans to build a thirty-bed hospital which will use a combination of modern and traditional Tibetan methods of healthcare. There are daily meditation classes and seminars and camps which offer a combination of meditation and trekking. The centre collaborates closely with the Students' Educational and Cultural Movement of Ladakh and has jointly sponsored a hydro-electric project at the historic Hemis monastery 40 kilometres out of Leh. They are also working on a solar energy project.

In 1962 the arrival of the Indian army in Ladakh to protect the region from hostilities from neighbouring countries heralded the beginning of an irreversible process of environmental decline. A decade later the Indian Government opened the area to tourists, and began to create a western-style infrastructure of roads, external dependencies and an accompanying cash economy. Large quantities of wheat, rice, firewood and coke at subsidized prices were ferried across the mountains by armies of trucks, accompanied by increasing numbers of tourists in jeeps and buses. Levels of air pollution have risen enormously, with attendant problems of bronchial disorders, aggravated by the need to inhale larger quantities of air at high altitudes.

Not only has traffic increased on the roads, leading to bigger demands for fuel and accompanying pollution, but the overall energy needs of Ladakh have risen so much that power currently represents the largest expense in the local government's budget. Millions of rupees have been spent over a period of twenty years to complete a 4-megawatt hydro-electric plant on the Indus River.

A detailed analysis of the problems created by modernization in Ladakh has been given by Helena Norberg-Hodge in *Ancient Futures*. Her forthright approach and suggestions for improvement deserve careful consideration. It is interesting, however, that in his preface, the Dalai Lama dwells on the inevitability of modernization and the need for adjustment over a period of time:

The abrupt changes that have taken place in Ladakh in recent decades are a reflection of a global trend. As our world grows smaller, previously isolated peoples are inevitably being brought into the greater human family. Naturally, adjustment takes time, in the course of which things are bound to change.[17]

In 1983 Helena Norberg-Hodge and her associates launched the Ladakh Ecological Development Group, with the aims of sustaining traditional culture and promoting sound development. Over a period of time the project has introduced solar ovens and water heaters, greenhouses to extend the growing season, hydraulic ram pumps (as at the Matho monastery) and micro hydroelectricity generators:

Our work aims to involve and train villagers to use technologies which augment rather than undermine traditional lifestyles and values. This has been accompanied by education and handcrafts programmes.[18]

Although the work of the Ladakh Ecological Development Group shows the influence of Buddhist ideas, its recommendations are expressed in a sufficiently ecumenical manner to appeal to other communities in Ladakh – Muslims and Christians – as well as Buddhists. An even more forthright analysis of Ladakh's environmental problems was offered by lecturers at the recently constructed government degree college outside Leh. Following a brief seminar in the staffroom, one science lecturer summed up the main problems in the region:

Sewage, garbage (especially polythene bags) and inadequate toilets (including hotels). There are large electricity shortages and vehicle pollution is especially serious at high altitudes. There are too few trees in Ladakh. People plant more and the rains increase; then the mud roofs of houses leak and people oppose the extra trees. They say that wooden roofs are too expensive. There is plenty of water in the Indus but irrigation is insufficient. The government does nothing. People need educating about the environment, and environmental education should be introduced into every curriculum.[19]

Not all these problems are the result of rapid modernization, and it is simplistic to argue that Ladakh should have been left completely alone to bask in its idyllic past. The Dalai Lama's view that things are bound to change is more realistic. However, granted that many things have changed both for better and for worse, what matters now is that the pressing problems currently facing Ladakh should be resolved with whatever resources are available. There is little doubt that the resources of religion, whether traditional Buddhism or an inclusive contemporary restatement of it which is acceptable to Muslims and Christians, have an important part to play.

Bhutan

The earliest residents of Bhutan migrated into the eastern region from northeast India and Burma. Sometime during the eighth century CE Tibetans migrated into Bhutan from the north, bringing their distinctive brand of Buddhism with them; their advent is marked by the legendary flight of Guru Rimpoche from Tibet into the Paro Valley on the back of a tigress. There are sound reasons to believe that he arrived in 747 CE, though his mode of travel is less certain. A third group of Nepalese origin migrated into the south of Bhutan at the beginning of the twentieth century in search of agricultural land. These three indigenous groups are collectively known as the Drukpa, and although they use a variety of languages and dialects among themselves, they increasingly share the national language, known as Dzongkha, which is compulsory in all schools.

Bhutan's present system of religious and secular government was set up early in the seventeenth century by Shabdrung Namgyal, a Tibetan lama of the Drukpa School. He unified the country and constructed a series of fortress monasteries known as *dzongs* in the major valleys. Administrative regions were established and an intricate legal system was codified. Following Shabdrung's death in 1651 there were intermittent civil wars until 1907, when Ugyen Wangchuck was elected the first king of Bhutan by a national assembly of monks, civil servants and regional representatives.

Bhutanese agriculture is mostly based either on lowland paddy fields where wet rice is grown, or upland dry fields where non-irrigated crops are to be found. These tend to be maize and barley in the eastern and central areas, with a limited amount of wheat plus potatoes, fruit and red chillies which blaze from the roofs of every available building at certain times of the year. The cultivation of rice in the lowlands is complemented by forests and grasslands which supply green manure, fuel and materials for buildings and agricultural implements. They also provide pasture for draught animals and space for water storage for irrigation and flood control.

Rice is not only grown in the low-lying south, where climatic conditions are especially favourable, but is also to be found in valleys where terracing and irrigation are much more difficult. Rice has always been a vital part of Bhutan's economy; it was the basis of the traditional barter system, and all Bhutanese festivals have some or other connection to this supreme offering to the gods.

Excluding the northern areas which are permanently under snow, 74 per cent of Bhutan is forested. In relation to population the forested area per person is 3.52 hectares, which is higher than in any other Himalayan region (0.84 in Nepal, 0.36 in Sikkim and 0.05 in Tehri/Kumaun; Ladakh is a desert area and cannot be compared). The main forest areas are among the inner Himalayan parts of eastern and central Bhutan and in the southern hills.

Bhutanese forestry began in earnest in 1959 with the formation of the

Forest Directorate. The Forest Department was set up in 1967, with responsibility for forest demarcation, management, silviculture, roads and afforestation. Commercial logging is carefully controlled, to such an extent that it is illegal to transport timber at night. The felling of trees was completely nationalized in 1979. In 1995 the Forest and Nature Conservation Act expanded and strengthened earlier legislation. The provisions of the Act include protected areas where human activities are regulated, social forestry and the preservation of endangered species. These are Bhutan's major contributions to Agenda 21, the United Nations' follow-up to the Earth Summit.

Ninety-seven per cent of Bhutan's total energy consumption is supplied by fuelwood, which is used mainly for domestic cooking and heating. The demand per capita is almost twice as high in the north as in the south and the total is twice as much as the national per capita average in Nepal. Commercial energy is derived from petroleum (60 per cent), coal (30 per cent) and hydro-electricity (10 per cent).[20] Phuntsholing on the southern border with India is the main commercial and industrial centre.

Bhutan requires 50 megawatts (MW) of electricity, which is more than satisfied by the Chukha hydro-electric station, a showpiece of efficient and environmentally sensitive planning. Situated on the main Phuntsholing to Thimpu road, the Chukha project diverts the waters of the River Wong Chu along a 6.5-kilometre displacement tunnel into an underground power station containing four 84 MW turbines. The resulting output, which averages 336 MW, is mostly sold to West Bengal, Bihar and Assam. Because the water is displaced rather than stored behind a large dam, the project avoids many of the acute environmental problems associated with big dams (e.g. the Narmada Dam).

We saw at the beginning of this chapter the ease with which Bhutanese Buddhism relates to contemporary social and environmental issues. In spite of its occult tendencies, *tantra* is essentially practical and this-worldly, emphasizing the interrelatedness of the various connections between human and non-human life. As Om Pradhan put it, it 'permeates all strands of secular life'. Many of Bhutan's 5,000 or so monks belong to the Drukpa branch of the Kagyupa Red Cap Order (though Om Pradhan has a Nyingmapa temple in the grounds of his house). They read the scriptures at weddings and funerals, treat illnesses, expel spirits, read horoscopes and perform masked dances at major festivals. In return – and also because almsgiving is considered meritorious – the people support them. Monasteries and nunneries are common throughout Bhutan.

On major religious and seasonal festivals the people and monks together congregate in the great *dzongs*. One of the largest of these is at Punakha, situated at the confluence of the Pho Chu and Mo Chu rivers. Passing through the gate of this massive fortress, one enters into a hive of Buddhist activity in which monks of every age are industriously rebuilding the central structures of the *dzong*. They work as labourers, artists, sculptors and architects, each contributing their personal effort. Watching them reverently and painstakingly at work on this beautiful and intricate edifice, which combines the historic legacy of the past with hope and optimism for the future, one cannot doubt that the Buddhist faith has an immense contribution to make to the social and environmental well-being of Bhutan.

———ややや———

The three major stages of resource use which provided a convenient background for my survey of the relationship between ecology and the early Indian tradition were complete by the time of the Buddha. Unlike the Hindu tradition, with its gods and a range of spiritual entities, Buddhism began as an analysis of human existence with implications for society, and as a philosophy with no need of theistic beliefs or sanctions derived from them, yet tolerant of them.

The Buddha's emphasis on desire, craving, attachment etc., and his practical measures for overcoming them, have enormous potential for the removal of the human causes of environmental degradation. Buddhist teaching, though anthropocentric in focus, attacks the root causes of human attitudes that account for so many of our modern social and environmental ills. Consumerism, materialism, wasteful competition, and the self-centred individualism that destroys the possibility of sound social and ecological relationships, are all challenged by the Four Noble Truths and the Noble Eightfold Path.

Together with the Buddha's central message, there are several important features of early Buddhism which possess continuing environmental significance. Buddhist monasticism is based on a frugal, communal lifestyle, and the earliest monastic treatises (e.g. the Pāṭimokkha) contain specific injunctions against the unnecessary destruction of plants and animals. Respect for animals is inculcated in the Jātaka narratives, and elsewhere. The Buddha's acknowledgement of spirits indicates that he respected the sanctity of trees and other natural objects that spirits inhabit.

In addition to such enduring features of early Buddhism that possess ecological significance, there have also been some important transformations. Although Ashoka's cosmic role as turner of the wheel of *dharma* contained Buddhist elements, I have described his self-understanding as founded on a stock of Indo-Aryan politico-moral ideas based on both Buddhism and the Hindu tradition. However it originated, it represents a potent transformation of tradition which played a vital role in the spread of civic Buddhism and is proving to be a powerful vehicle for social and environmental improvement in south and southeast Asia.

The theory of interdependent co-arising (*paṭicca samuppāda*) is foundational for both Theravādins and Mahāyānists, and there is no evidence to suggest that modern science is weakening its credibility (as in the case of the corresponding Hindu belief).[21] It embraces all life and the natural world, and is therefore highly charged with ecological significance. In the next chapter we shall see how it has been imaginatively transformed by Buddhadāsa to encapsulate the view that 'the world is a conjoint, interdynamic, cooperative whole'.[22]

Śūnyatā also provides a fruitful focus for ecological concern. Since all phenomena are characterized by 'emptiness', then all are equivalent, and there can therefore be no ultimate hierarchy among various life forms or between life and non-life. Helena Norberg-Hodge's interpretation of *śūnyatā*, based on her research in Ladakh, as the view that all things ultimately dissolve into a web of relationships, is a creative interpretation of the concept. Vajrayāna Buddhism and associated *tantric* schools give meaning and purpose to human activities by stressing enlightenment here and now, and of all the Mahāyāna's potent beliefs, perhaps the most ecologically charged is that of the *bodhisattva* who refuses to enter *nirvāṇa* in order to work for the salvation of all living beings – even the last blade of grass.

6   Thailand: a case study

A train journey and a short bus ride brought me to the path leading up to the monastery where a monk famous for his knowledge of herbal medicines in Thailand resides. There was no path beyond a few metres, however, and the young, yellow-robed novice who had been waiting for me skipped bare-footed from boulder to boulder higher and higher up what turned out to be a small mountain.

By the time the monastery came in view I was so exhausted that I could hardly breathe. Catching sight of the abbot, I *wai*-ed him respectfully, and collapsed at his feet.[1] Temple lay assistants plied me with glasses of a herbal cure for sunstroke, but to no avail. At a gesture from the abbot the novice vanished into the trees, to return after a few minutes with a strip of paracetamol. The combination of treatments worked, and I was soon inspecting and photographing the abbot's collection of medicinal plants in the monastery precincts.

The results of this research programme are summarized in Appendix A (pp. 176–180), and the locations of the monasteries visited are shown on the map of Thailand (Figure 6.1, p. 87). The significance of this study is that it highlights a traditional repository of knowledge which has become increasingly important as we begin to understand the extent of global biodiversity loss. Not only are the medicinal plants (Thai: *samun prai*) important in their own right, but their cultivation by Buddhist monks represents the transformation of a practice enshrined in the religious texts, which has enormous contemporary significance as we contemplate our battered global ecosystem. We shall return to this issue later.

During the past two centuries the southeast Asian region has experienced enormous economic and environmental changes. These occurred initially during the colonial era, but have continued with even greater momentum since the end of the Second World War. The demands of expanding empires created a voracious appetite for timber for buildings, railway sleepers, boats and fuel. However, the more sustained pressure for permanent agriculture based on plantations has resulted in an even greater level of deforestation, as each country in the region has become more firmly integrated into the global capitalist economy.[2]

Colonial rule in southeast Asia was accompanied by the introduction of

Figure 6.1 Thailand: fieldwork locations

European science and technology. This was to some extent a consequence of the need to understand natural resources in order to extract them. It was also the result of new technologies associated with the railway, the steamship, the telegraph and mechanical means for extracting minerals such as petroleum. Traditional methods of resource use and cultivation were increasingly replaced by the allegedly more scientific approaches of the colonists and their post-colonial successors. In recent years it has become clear that both the colonial and post-colonial eras of resource extraction have wreaked a terrible environmental toll throughout southeast Asia, much of it in the name of 'progress', 'science', and even 'development'. It is also becoming increasingly apparent that some of the traditional patterns of resource use are far better suited to long-term social and environmental sustainability than the ones which have replaced them.

In this chapter we shall consider ways in which Buddhism can address social and environmental problems in Thailand. Following a review of some of the environmental parameters of the country, I shall summarize the history of Buddhism through various phases into the nineteenth and twentieth centuries, when it underwent a number of reforms under King Mongkut, and as a result of initiatives by progressive monks. As in previous chapters we shall identify some features of contemporary Thai Buddhism which are of ecological significance and which reflect continuity with tradition, and others which represent a significant transformation of the past.

Environmental parameters

Thailand is part of a transitional region between moist evergreen and dry deciduous forests. Subtypes of forest include conifers, hill and lowland evergreens and mangrove, varying according to climate, altitude and soil constituency. Between the late 1940s and the end of the 1980s, Thailand's forest cover declined from 69 per cent to 15 per cent, a sad reflection on the ability of successive governments to stem the tide of exploitation. Although Thailand was never formally colonized – largely on account of adept political footwork by two nineteenth-century kings – the profits from the international demand for timber proved as irresistible to the Thai as to their neighbours.

Deforestation in the northern highlands of Thailand has adversely influenced major water flow systems, with the result that the rice economy of the central regions has been seriously impaired. Loss of trees has caused soil erosion and increased sedimentation in rivers, reservoirs have become silted, and in some areas there has been drought. Without sufficient forest cover to maintain the ecological balance of the north and adequate flows of water to irrigate what has been described as the rice bowl of the kingdom, the overall situation has become extremely serious.

During the last thirty years there has arisen a significant 'green' lobby which the politicians have started to take seriously, in addition to acknowledging themselves the magnitude of forest cover loss, flooding, etc. According to Jonathan Rigg, its growth is also related to a highly charged political incident in 1973, when soldiers involved in the suppression of student demonstrations wantonly killed a number of animals.[3] At about the same time there were protests against pollution from sugar mills affecting the Mae Khlong River. Both these occurrences took place at a time when the international community was beginning to heed the warnings of the 1972 UN Conference on the Human Environment in Stockholm.

In 1988 plans to build a dam in Kanchanaburi province across the larger tributary of the River Kwai met with such sustained opposition that the government abandoned the idea. Opposition included local groups, farmers, students and Buddhist monks. At the end of 1988 severe floods in the southern province of Chumphon caused more than 300 deaths. The severity of the floods was attributed to deforestation in the surrounding hills, and it was largely as a result of public concern about this disaster that a national ban on logging was imposed in 1989.

Public unease was responsible for the cancellation in 1992 of an inadequate plan to redistribute land for the poor living in forest reserves, and the following year the discovery of a discharge of untreated waste led to the closure of a pulp and paper factory. Although the government has been slow to respond to public anxiety about such matters, it did produce a four-year plan in 1992 which included several important environmental measures: endorsement of the 'polluter pays' principle, for example. In spite of such progressive legislation, however, environmental issues tend to be addressed from the perspective of which political group is in favour of what, rather than according to urgency.

From about 1984 onwards the Thai government has promoted reforestation projects, but these have led to the expulsion of farmers from land which has then been used to set up commercial eucalyptus plantations to supply paper mills. The farmers were offered resettlement on land which they discovered on arrival to be occupied. They were then obliged either to move illegally into virgin forests or to migrate to the cities – usually Bangkok. Not surprisingly, many of them have taken part in demonstrations to obstruct bulldozers and destroy eucalyptus seedlings. Some of the land taken from farmers has been used for the construction of hotel complexes and golf courses (high-class Thai love golf!).

Less than 17 per cent of the cultivated area of Thailand is irrigated and most farmers therefore depend on rain. In 1992 a severe drought led to a spate of irrigation projects and dams which flooded areas of forest, thus exacerbating the tensions caused by compulsory resettlement. Several large dams such as the Pak Mun Dam in the northeast (which is supported by the World Bank), have been vigorously opposed by local groups who believe they are being constructed to supply industry with power and water, rather than to improve water supplies for agriculture. Other serious water-related problems include eutrophication caused by the dumping of liquid manure from livestock rearing, nitrification resulting from the excessive use of fertilizers, the pollution of rice paddies caused by effluent, and the destruction of fish stocks due to untreated waste water from wood pulp factories and sugar mills.

The association between water, rice and prosperity has been acknowledged in Thailand through the centuries. It is summed up in the well-known inscription of King Ramkhamhaeng, dated 1292: 'In the time of King Ramkhamhaeng, this land of Sukhothai is thriving. In the water there is fish, in the fields there is rice.' The social and environmental problems of the last few decades now threaten to sever irreparably this vital link.

Ashoka's legacy

There are two main sources of Thailand's monastic tradition: the Indian legacy of Ashoka Maurya, and the southeast Asian elements which derive

90   *Religion and Ecology in India and Southeast Asia*

from the first and second centuries of the Common Era. The latter are complex and include Hindu, animist and both Theravāda and Mahāyāna components, but it is the emerging notion of kingship which gives the tradition its cohesion.

Ashoka's exalted view of himself as *cakravartin*, or turner of the wheel of cosmic law, seems to have been derived from a range of Indo-Aryan politico-moral ideals, in part Buddhist. There is no trace in Ashoka's edicts of the notion of the king as *bodhisattva*, or future cosmic liberator, which first appears in Burma and Siam. Another major influence in southeast Asia was the cult of the king as *devarāja* (god-king); this came from the Indianized Khmer and Mon kingdoms which made use of brahmins at court rituals and believed in a cosmology modelled on parallels between the superhuman macrocosmos and the human microcosmos. There were also animistic beliefs in spirits.

The oldest Buddhist site in Thailand – at Nakhon Pathom – dates from the sixth century CE, but there is evidence of Buddhist activity before then. The earliest monastic communities were organized under a teacher, even though the Buddha himself appointed no successor and advocated a form of democracy. The Pāṭimokkha was recited bi-monthly, procedures were adopted for resolving disputes, and special rules were generated for corporate life during the rainy season, when monks may not travel without permission from the abbot. Initiation required at least ten monks, an *upadhyāya* (ordainer) and an *ācārya* (elder teacher).

From the beginning of the twelfth century onwards the Siamese polity began to emerge as an independent entity. It was based on Sukhodaya and strongly influenced by both Burmese and Sinhalese political and religious elements. The Sukhodaya kingdom is usually dated from 1257 until 1350, when the power of the reigning monarch had been largely superseded by the king of Ayutthaya. The Ayutthaya kingdom was twice invaded by the Burmese. On the second occasion, in 1767, Ayutthaya was destroyed, and after a period of political chaos a new national capital was built in what became the twin cities of Thonburi and Bangkok. Subsequent kings of the Chakri dynasty were all known as Rama, and the first of these, Rama I, set about reforming the *saṅgha* with vigour. In 1801 he ordered the expulsion of 128 monks on the grounds that 'they had been guilty of all kinds of ignoble behaviour, namely drinking, wandering about at night, rubbing shoulders with women, using improper language, [and] buying silly things from Chinese junks'.[4]

Tambiah has characterized the Thai polity as 'galactic', since the king's domain in his capital city was replicated in more or less independent regional centres whose governors were appointed by him.[5] During the nineteenth and twentieth centuries this galactic polity has become increasingly replaced by a 'radial' one, in which the capital acts as a magnet for the entire country.

An important consequence of the radial polity is the process whereby monks gravitate to the metropolis via a network of elaborate monastic routes. Many of them from poor areas in the provinces ordain as novices in their early teens to obtain the best grades open to them in their home area, subsequently moving to the nearest provincial capital to take advantage of the better educational opportunities. They become monks in their early twenties and usually remain in the *sangha* for many years, either disrobing in their mid-thirties to raise a family or continuing into old age. However, as free government schooling improves in rural areas, the educational motive for ordination is declining. It is also quite common for the sons of the well-off urban classes to ordain for a few weeks during the rainy season (*phansā* or Lent). Such short-term ordinations are undertaken to bestow merit on parents or as a rite of passage to adulthood.

Monastic reforms

Although never formally colonized, the Thai economy during the nineteenth century possessed many features in common with colonial states. Industrial development remained minimal while revenue was derived mainly from the export of agricultural products such as teak and rice plus tin and various minerals. The kings of this period were able to utilize their wealth to introduce large-scale administrative reforms.

Prince Mongkut, who became Rama IV in 1851, had already been a monk, and it is therefore not surprising that he made a number of major changes to the *sangha*. Dissatisfied with the way of life of the forest-dwelling monks, he insisted that primacy be given to learning. He inspired and led a movement within the *sangha* known as *dhammayuttika*, meaning 'those who adhere to the law (i.e. scripture)'. *Dhammayuttika* monks wear their robes across both shoulders and generally observe the monastic Vinaya rules more closely than the unreformed *maha nikai* monks. Although these two groups of monks are distinct and live in separate *wats* (monasteries or temples), it is not strictly correct to refer to them as sects as there are no major doctrinal divergencies.

Mongkut was also responsible for the revision of the Pali canon in collaboration with Sinhalese scholars. He rejected much of the fantastic cosmology adhered to by his predecessors, and adopted a rational and scientific approach to religious dogma. Through his advocacy of science and reason, Mongkut paved the way for this-worldly interpretations of Buddhism by scholars such as Buddhadāsa and the more recent socially and environmentally active monks.

King Chulalongkorn (Rama V) continued Mongkut's policies of promoting education, and in 1898 entrusted the *sangha* with the implementation of a national programme of primary education. In 1902 he passed the first of three *Sangha* Acts, which set out the obligation of monks to observe 'three types of laws: the law of the land, the Vinaya and custom'. The Act set out the duties of various monastic authorities and standardized *sangha* administration. Following the replacement of absolute by constitutional monarchy in 1932, the second *Sangha* Act, of 1941, was a move in the direction of ecclesiastical democracy, while the third, in 1963, reflected the authoritarian policies of Prime Minister Sarit. It concentrated power in the person of the supreme patriarch and replaced various *saṅgha* committees with a council of elders known as the *Mahatherasamakhom*. Although Thai society has subsequently become more democratic, the authoritarian tendencies which Sarit imposed on the *saṅgha* have never been revoked.

Throughout the nineteenth and early twentieth centuries Thai society experienced an exposure to secularization comparable to that of India. If it ultimately proved to have fewer consequences, then this was because of the absence of formal colonization and because European languages did not replace Thai as the medium of secondary education.

During the late nineteenth and twentieth centuries we can trace various responses to secularization, whereby Buddhist tradition was reasserted, adapted or rejected. Mongkut himself, although clearly a reformer, is probably best considered as an agent of secularization somewhat similar to Ram Mohan Roy, rather than as an adapter of the tradition. In discussing responses to secularization, Tambiah adds the category of neo-traditionalism, 'an ideology designed to keep change to a minimum and defend the status quo as far as possible'.[6] This does not fit into the more fundamental and logically coherent three-category scheme which seemed to work well in the case of India, and we shall not consider it further.

During the 1930s there were discussions among educated Thai laymen about the implications of science for religious belief. Arising out of these discussions some monks and former monks put forward reassertive traditional views which challenged science. Luang Vichitr Vadhakarn, for example, insisted that the Buddha used *iddhi* (miraculous powers) to demonstrate his superiority over the Hindu *yogins*. An ex-monk and Pali scholar called Nai Pui Saeng claimed that traditional Jātaka teaching about heaven and hell should be maintained in the face of scientific opposition. These were fairly typical of the reasserters of religious tradition, although unlike Dayananda Sarasvati, they did not have any broad-based lower middle-class following.

The most interesting responses to secularization were by those who adapted their religious tradition, often borrowing from the scientific ideas of the West. In Thailand this category is best represented by Buddhadāsa, whose this-worldly reinterpretation of Buddhism will be considered in detail.

Buddhadāsa Bhikkhu

Buddhadāsa Bhikkhu ('the monk who is a servant of the Buddha') was born in 1906 in Chaiya in southern Thailand. His mother was Thai, his father was a second-generation Chinese businessman, and his original name was Ngeuam Panich. He describes the primary influences on his life as his mother, the local *wat* and nature.

At the age of ten the young Buddhadāsa (Thai: Putatāt) became a *dek wat* (temple assistant) for three years, during which time he learned to read and write. He became familiar with temple life and collected medicinal herbs for the abbot. He ordained as a monk at the age of twenty and went to Bangkok to undertake Pali studies, but he was unhappy with city life and returned to Chaiya where he founded Suan Mokkhabalārāma ('the garden of the power of liberation') in 1932. The originality of his views soon began to attract attention, and many distinguished visitors came to see him. He gave lectures in Bangkok, teaching in a rational and incisive manner which appealed to the educated middle classes. He died in July 1993.

The cornerstone of Buddhadāsa's beliefs is that only emptiness or the void (*śūnyatā*) truly exists; everything else has a qualified reality – a view with strong similarities to the Mādhyamika philosophy from which the Mahāyāna stream of Buddhism developed. All existence is composed of transitory, impermanent events, but *śūnyatā* never changes; it is absolute being, absolute truth, *nirvāṇa* and the body of essence of the Buddha.

If ultimate reality is unchanging, then the cycle of rebirth (*saṁsāra*) cannot be a temporal process leading to *nirvāṇa*; it must be here and now, like *nirvāṇa* itself. Consistent with this interpretation, *anattā* (no-self) can now be regarded as a statement about the removal of self-centredness – the cause of attachment and consequent suffering – by wholesome actions. Wholesome actions and their consequences are determined by *paṭicca samuppāda*, or interdependent co-arising, which embraces all life and non-life in a web of interdependence.

Buddhadāsa's this-worldly 'here and now' interpretation of cardinal Buddhist doctrines is particularly appealing to busy professionals who have no time to meditate, and feel inferior to those who have. For them, *śūnyatā* represents a quality of activity in which the mind is tranquil, integrated and non-attached, as in meditation, though not through any meditative technique. In a personal interview with Buddhadāsa at Suan Mōkh in 1974, I asked him how similar this is to *niṣkāma* karma (work without attachment to its fruits) in the Bhagavadgītā.[7] He acknowledged the similarity, but pointed out that in Buddhism the 'empty' or 'void' mind becomes progressively '*nibbāned*', or cooled, which is the etymological meaning of *nibbāna* (in a more temperate climate our hearts might be 'warmed'!).

Mahāyāna influence on Buddhadāsa is variously attributed to his Chinese father and the fact that the Mahāyāna tradition has always been strong in southern Thailand. The museum at Suan Mōkh contains many illustrations of *bodhisattvas*. There is an oil painting of the leaders of the world's major religions greeting one another, and a Zen cartoon of a group of people fastened together by a cord; the caption reads 'Human arrangement – by flowers'.

Buddhadāsa was too shrewd to become actively involved in politics, either to the right or to the left. He often spoke about '*dhammic* socialism', however, and maintained that there can be no ultimate separation between the spiritual and the social. *Dharma* (Pali: *dhamma*) is one with *dhammajāti*, or nature, which is the sum total of reality: 'The trees can speak, the rocks can speak, the pebbles and sand, the ants and insects, everything is able to speak.'[8]

Donald Swearer has studied the later years of Buddhadāsa's life, the period in which he became increasingly concerned about the destruction of the natural environment. According to Swearer, the elements of Buddhadāsa's ecological hermeneutic are to be found in his lecture at Suan Mōkh in 1990, in which the two central terms are 'care' (Pali: *anurakkhā*; Thai: *anurak*) and 'nature' (Pali: *dhammajāti*; Thai: *thamachāt*). The first of these terms is often loosely translated into English as 'conservation', and monks who oppose tree felling are described as 'forest conservation monks' (*phra kānanurak pā*). However, when Buddhadāsa uses this term he gives it a much more empathetic significance:

> One cares for the forest because one empathizes with the forest just as one cares for people, including oneself, because one has become empathetic. *Anurak*, the active expression of a state of empathy, is fundamentally linked to non-attachment or liberation from preoccupation with self, which is at the very core of Buddhadāsa's thought ... It is just such non-attachment or self-forgetting – the heart of the *dhamma* – that we learn from nature. We truly care for our total environment, including our fellow human beings, only when we have overcome selfishness and those qualities which empower it: desire, greed, hatred ... Caring in Buddhadāsa's *dhammic* sense, therefore, is the active expression of our empathetic identification with all life forms: sentient and non-sentient, human beings and nature.

Caring in this deeper sense of the meaning of *anurak* goes beyond the well-publicized strategies to protect and conserve the forest, such as ordaining trees, implemented by the conservation monks, as important as these strategies have become in Thailand. This is where the second term, *thamachāt*, enters the picture. The Thai term *thamachāt* is usually translated as 'nature'. Its Pali root, however, denotes everything that is linked to *dhamma* or that is *dhamma* originated (*jāti*). That is to say, *thamachāt* includes all things in their true, natural state, a condition that Buddhadāsa refers to as 'norm-al' or 'norm-ative' (*pakati*), that is, the way things are in the true, dhammic condition. To conserve (*anurak*) nature (*thamachāt*), therefore, translates as having at the core of one's very being the quality of empathetic caring for all things in the world in their natural conditions; that is to say, to care for them as they really are rather than as I might benefit from them or as I might like them to be. Indeed, *anurak thamachāt* implies that the 'I' is not over against nature but interactively co-dependent with it. In other words, the moral/spiritual quality of non-attachment or self-forgetfulness necessarily implies the ontological realization of interdependent co-arising.

From an ethical perspective this means that our care for nature derives from an ingrained selfless, empathetic response. It is not motivated by a need to satisfy our own pleasures as, say, in the maintenance of a beautiful garden or even by the admirable goal of conserving nature for our own physical and spiritual well-being or for the benefit of future generations.⁹

Buddhadāsa's view that *dharma* embraces the social and environmental spheres of human activities has inspired a number of progressive monks, though it is important to recognize that he did not encourage them to undertake community development activities themselves because this was the role of lay people. Among his chief followers are Phra Depvisuddhimedhi (also known as Phra Paññānantha), abbot of the Wat Cholapratan Rangsarit, and Phra Payom Kallayano, abbot of the Wat Suan Kaew (both in Nonthaburi). Phra Payom, who was a student of Buddhadāsa for seven years, spices his sermons with street-level slang and is very popular among young people. A typical off-the-cuff comment to me about monks' development programmes was as follows: 'If a monk is concerned about how to solve social problems, then he will not have time to think about removing any skirts.'

Phra Prayudh Payutto (also known by various honorific names such as Phra (or Chao Khun) Rajavaramuni and Phra Dhammapiṭaka) is a great admirer of Buddhadāsa, but differs from him in several important respects. His monastic career followed the traditional trajectory of Pali studies and the position of deputy secretary-general of Mahachulalongkorn Buddhist University, where he pioneered development activities for young scholar monks. Donald Swearer contrasts the environmental views of Phra Prayudh and Buddhadāsa as follows:

> Phra Prayudh grounds his argument for the value of nature for religious practice in stories of the Buddha and the early disciplines found in Pali texts. Buddhadāsa also links nature and religious practice to spiritual realization but does so by using Suan Mōkh as his primary illustration rather than citing specific passages in canon and commentary. Phra Prayudh, furthermore, makes a strong appeal to reason. Unlike some Thai Buddhist environmentalists who encourage such practices as ordaining trees or the promotion of a tree deity cult to preserve a stand of trees, Phra Prayudh believes that modern Buddhists need to go beyond appealing to Buddhist values, such as gratitude and loving-kindness, and citing scripturally grounded stories of the Buddha and the early *saṅgha*, and should utilize scientific evidence to address global problems, such as pollution and environmental preservation.¹⁰

Tambiah attributes the resilience and creativity of Thai monasticism to the absence of colonization combined with the intrinsic character of Buddhism:

> Virtually at all levels of society the integral relevance of their religion for conduct is not in doubt ... This attitude is partly the result of the greater sense of intactness and continuity experienced by the Thai as compared with other Asian societies actually colonized by Western imperial powers.

But it also derives from the intrinsic character of Buddhism itself – how its tenets relate, on the one hand, to the confident claims of positivist science and, on the other, to the concerns of the politico-social order.[11]

Buddhadāsa and, to varying degrees, other monks, have been able to straddle both these worlds.

Towards the end of his life Buddhadāsa became seriously ill. Everybody, including King Adulyadej Bhumibol (Rama IX), became deeply concerned, and Thailand's best doctors were despatched to attend to him. Three choices were open to them: they could treat him at Suan Mōkh, move him to a nearby hospital, or bring him to the Sirirath Hospital in Bangkok, one of the best in southeast Asia. The hospital's director, Dr Prawase Wasi, a long-standing admirer of the monk, was asked to convey a message to him from the King, requesting him 'not to leave his body so that he can help to maintain the *sasana* (religion)'.

'You can ask,' responded Buddhadāsa, 'but it all depends on causal conditions. If there are factors that enable the body to live, it will. If not, it won't. Don't try to carry the body away to escape death.'[12] He remained at Suan Mōkh, and recovered. He died two years later, aged 87.

Social and environmental activities

In 1964 monks living in Bangkok began to take part in two kinds of outreach to the provinces. The Phra Dhammatuta programme started that year under the auspices of the Department of Religious Affairs (part of the Ministry of Education and run mostly by former monks). Its primary aim was to promote national integration by strengthening people's attachment to Buddhism. It had a strong missionary emphasis and has sent a number of monks each year to work abroad.

The Phra Dhammajarik ('wandering *dhamma*') programme was jointly sponsored by the Ministry of the Interior via the Department of Public Welfare and the *sangha* in 1965; its main thrust was to spread Buddhism among border and tribal people. Somboon Suksamran describes it as 'a kind of moral rearmament mission to the northern areas where the tribesmen live and have been threatened by subversion'.[13] Both these programmes had strong political overtones from the outset and were not particularly successful.

Independent development schemes, known as *dhammapatana* ('development through *dhamma*'), have been organized by individual *wats* such as the Wat Phra Singh in Chiang Mai and by the two Buddhist universities in Bangkok. The first training programme for monks at Mahachulalongkorn Buddhist University began in 1966, and by 1972 was sending about sixty trainees a year to work in the provinces.[14] The programme also brought monks from the provinces to attend two-month courses in Bangkok; these included training in community and rural development, sanitation and first aid, and public health. There was also a lecture course on ecology and the environment. Mahamakut Buddhist University, which is mainly for *dhammayuttika* monks, ran similar programmes, initially jointly with Mahachulalongkorn.[15] After a period of training, monks went to poor provincial areas where they took part in schemes to construct roads, bridges, school and temple buildings, wells and toilets. They also helped to install water pumps and electricity lines. William Klausner has described the involvement of monks in these activities in northeast Thailand in graphic detail.[16]

Monks sometimes operate their own development programmes based on particular *wats*. In Chiang Mai province Phra Khru Mongkol Silawongs, abbot of the Wat Bupparam in Chiang Mai city, runs vocational training schemes for electricians, builders and architects. Monks are taught how to help hill tribespeople to dig wells and build roads, and women are trained in weaving, sewing, toy making, and fruit and flower growing. In an interview Phra Khru Mongkol refuted the notion that his development work was political or that he wanted to make converts to Buddhism. He did it for its own sake and because the Buddha taught his followers to work for the welfare and happiness of others.[17] Chao Khun Rajavinayaporn is a senior *dhammayuttika* monk (Phra Khru Mongkol is *maha nikai*), who is deputy abbot of the Wat Chedi Luang in Chiang Mai. He supervises schemes to train women, and helps hill tribes raise money for development work, which includes the use of biogas and improved cooking facilities.[18]

The involvement of monks in social and environmental programmes raises important questions about appropriate and inappropriate behaviour for a monk. According to the Pāṭimokkha (which is part of the monastic Vinaya) a monk may not damage a plant or dig the earth (which might destroy small living creatures), but there is no reason why he cannot saw a log if somebody else has cut down a tree. The 227 rules of the Pāṭimokkha do not apply to a novice, who is subject only to ten precepts. The rules are particularly important to Thai monks because the Vinaya played a major part in King Mongkut's *saṅgha* reforms. Most Thai are therefore well aware of them, and public opinion is extremely sensitive to what is and what is not appropriate behaviour for a monk. From time to time the national press erupts with a story about some monk who is behaving inappropriately; this happened in July 1978 when Phra Kittiwuddho, a controversial politically rightist monk, was discovered to have had a Volvo car smuggled into the country for his use.[19]

Public opinion is less critical, however, if it is clear that there is a positive reason which can be endorsed from a Buddhist perspective why certain technically inappropriate activities are necessary. Thus, for example, Phra Chamrun Panchan, former abbot of the Wat Tham Krabok in Saraburi (he died in May 1999, to be succeeded by his brother), discovered a herbal medicine which, if administered in a therapeutic community based in his monastery, is highly successful in curing heroin and opium addicts.[20] Many of these addicts are teenagers, however, and it would hardly be reasonable to treat only the young men and not the women. The monks therefore feel obliged to do many things which, according to the Pāṭimokkha, bring them into an inappropriate amount of contact with women. They must also, from time to time, operate a sauna, clear up lay people's vomit and pursue absconders. However, all this technically inappropriate behaviour may be condoned because the Buddha preached against the use of intoxicants. A person who cures drug addicts is therefore doing what the Buddha would have approved.

In 1976 Dr Prawase Wasi, director of the Sirirath Hospital and an eminent haematologist, arranged a three-week course on healthcare for monks at the Wat Thongnoppakun in Thonburi. This was followed by shorter five-day courses for groups of up to fifty monks at other Bangkok *wats*. These courses included the prevention and diagnosis of illness, childcare, and the treatment of illness using inexpensive traditional and modern medicines.[21] I have conducted studies of these courses and their effectiveness, in one case with particular reference to the availability and use of herbal medicines. Appendix A gives a list of the medicinal plants I photographed at a number of *wats* and subsequently identified from a catalogue in the library of Chulalongkorn University in Bangkok. The catalogue, by Ratdawan Boonratanakornit and Thanomchit Supawita, is called *Names of Herbs and their Uses*. It bears no date, and it could not be found anywhere else.[22] Appendix A lists the uses of each plant as specified in the catalogue.

I have also conducted studies of healthcare possibilities for monks and *mae chii* (lay nuns) in urban situations.[23] These *maw phra* (doctor-monk) schemes have been highly successful and have done a great deal to enhance public respect for the *saṅgha* and its lay supporters. Monks who practise basic healthcare are often dubbed 'bare-headed doctors'.

The social and environmental programmes described so far have aroused little controversy, but more recent attempts by monks to stem the tide of deforestation have led to major confrontations with the authorities. Trees have been ordained by encircling them with saffron cloth to prevent them from being felled, and sacred groves have similarly been created with sacred thread. Phra Thui from the Wat Dong Sii Chomphuu in Sakon Nakhon province has protected trees in this manner, and Phra Prajak Khuttajitto, a monk living in Dongyai forest in Buriram province in the northeast, was sent to prison for encroaching on a forest designated for the felling of trees.

The imprisonment of Phra Prajak occurred because of his support for poor farmers who were being resettled to make room for eucalyptus reforestation under a controversial resettlement programme set up in 1990 by General Suchinda Kraprayoon. Critics discovered that the real reason beneath the official veneer of environmental concern was that the army wanted to make money from private plantation companies. Phra Prajak was arrested in April 1991 for encroaching on forest reserve land and again in September, but in July the following year a new prime minister, Anand Panyarachun, abolished the resettlement programme.

Many of the social and environmental activist monks receive encouragement from Sulak Sivaraksa, who coordinates what has come to be known as 'engaged Buddhism'. Sulak is a Sino-Thai originally from Bangkok, who graduated in law in Britain. Following a distinguished literary and publishing career, during which he wrote extensively about the renewal of society through Buddhism, he attracted international attention in 1991 as a result of reactions to his comments about the monarchy. His remarks, made during a lecture at Thammasat University, could be construed as criticisms of the King, and he was arrested on charges of *lèse-majesté*.

Sulak believes that international capitalism and the consumer culture are primarily responsible for undermining Siamese society (he does not like to call it Thai!):

> The great department stores or shopping complexes have now replaced our *wats*, which used to be our schools, museums, art galleries, recreation centres and cultural centres as well as our hospitals and spiritual theatres. The rich have become immensely rich, while the poor remain poor or even become much poorer . . . Not only our traditional culture, but our natural environment, too, is in crisis.[24]

Sulak's Buddhism owes much to Buddhadāsa. Thus he regards *nirvāṇa* not as a metaphysical reality but as an experience beyond the limits of the mundane: 'inner freedom, equilibrium, peace, void of angst and a sense of being entirely "at home" and unthreatened in the universe, which expresses itself both in a positive affective state and in compassion for all forms of life'.[25]

Sulak was eventually acquitted of the *lèse-majesté* charges, but was arrested in 1997 for taking part in direct action against the construction of the Yadana Gas pipeline in Kanchanaburi.[26] Environmental groups oppose this construction because it will destroy virgin forests in Thailand and encourage the military regime in Burma with its repressive policies for relocating villagers on the Burmese side of the border. Tree ordinations, attempts to lie down in front of trucks and appeals to the international community are all part of the campaign. The struggle continues . . .

However, Buddhism can be a vehicle for environmental improvement in less dramatic ways. In education, for example, there are experimental school programmes which utilize Buddhist principles to promote community-oriented energy- and resource-efficiency schemes. One of these is based at the Wat Tongpuboran-Khanissorn municipal primary school in Ayutthaya. It began in 1997 as a collaborative venture between the Ministry of Education, the National Energy Policy Office and the Thailand Environment Institute, and is known as the DAWN project.

According to Khru Surin, one of the school's teachers,

> Our attempts to jump onto the economic expressway have landed us in disaster. Look at the vast rice fields . . . In a rush to get rich quick,

farmers have poured in so many toxic chemicals to boost yields that they've killed the fish in the ponds as well as put their own health at risk.[27]

The project manager is Dr Uthai Dulyakasem, former dean of the Faculty of Education at Silpakorn University. He believes that all education must be geared to the Buddhist notion of right (or perfected) understanding (the first step on the Noble Eightfold Path):

> Education must lead towards Right Understanding, which in Buddhist teaching is *sammā-ditthi*. The future generations must realize that saving energy is not a personal matter but a vital concern of the entire society... The DAWN programme hopes to enable people to see themselves as responsible for the entire process, from production to consumption. This means that they must be able to see and connect things from a holistic perspective.[28]

A reform in the entire learning process is essential for the creation of a better future; it must change people's patterns of consumption while at the same time nurturing their sense of social responsibility. The school curriculum draws upon the resources of the local community and integrates environmental matters into every subject, using them to explore the links between academic disciplines usually taught in isolation from one another. Thus, for example, a discussion of how much energy has been utilized in the production, transportation and consumption of a commodity could form part of a course in geography, physics or mathematics, and a class on Buddhism could be used to analyse the manner in which commercial advertisements stimulate the desires and illusions of unbridled consumption.

Urban sects and movements

In addition to a range of social and environmental activities involving monks, there are several urban movements which display varying degrees of social commitment. I conducted a survey of these in 1993.[29] They were the Engaged Buddhists, centred around Sulak Sivaraksa and based in Bangkok, but with substantial rural support; the movement around Phra Kittiwuddho; Santi Asok; Buddhadāsa Bhikkhu and his disciple, Phra Payom Kallayano; and the Wat Phra Thammakaay movement. We shall also mention Samnak Paw Sawan and Hooppha Sawan, which were not included in the survey but have recently begun to take environmental issues seriously.

Kittiwuddho's activities are based on Cittabhavana College in Chonburi, which was founded under royal patronage in 1957. Novices and monks undergoing courses receive training in a wide range of skills, which include driving tractors and operating lathes and oxy-acetylene equipment. According to Kittiwuddho, 'monks are people in the world; they should help the world'.[30]

Kittiwuddho combines extreme rightist politics with imaginative programmes for social reform. He believes that social inequalities and environmental problems are the result of bad karma, but that Buddhists should do their best to ameliorate them. In 1987 he recommended the setting up of rice mills in temples, and supported General Chavalit Yongchaiyuth's tree-planting campaign in the northeast. His social and environmental engagement may be summed up as social improvement without participatory democracy.

Whereas Kittiwuddho and his supporters remain closely in touch with the centres of ecclesiastical and secular power, Santi Asok has always maintained strong roots in the northeast, though it became prominent in the capital in the mid-1980s when one of its most charismatic members, Chamlong Srimuang, became governor of Bangkok. In May 1992 Chamlong was thrust into the leadership of a popular revolution against prime minister General Suchinda Kraprayoon, a bloody struggle which was resolved only by the personal intervention of the King. Television audiences around the world were intrigued by dramatic photographs of the King, seated in informal attire, confronting the prime minister and the governor together, as both lay flat on their faces in front of him.

Phra Photirak, the leader of Santi Asok, was born into a Sino-Thai family in the northeast in 1934. Following his ordination to the monkhood in 1970, he became strongly influenced by the forest monks and the ascetic practices of Luang Paw Man. Santi Asok's northeastern origins and the influence of the forest monks have equipped it with an instinctive sensitivity to environmental issues. Vegetarianism and the simple ascetic lifestyle of its members have proved appealing to sections of the Bangkok public which have grown tired of the city's brash materialism. Ordained and lay followers at the sect's headquarters learn a variety of practical skills, such as computer programming and the construction of radio equipment. The sect ordains women as monks – one had been a famous film star – though such ordinations are illegal.

On the roof of Santi Asok's Bangkok headquarters sat a handsome teenage boy, resplendent in white robes, in front of an electric typewriter, doing his physics homework. A few feet away from him lay an open coffin containing a rotting corpse. The boy explained that this was to enable adherents to understand the meaning of life by meditating on the various stages of decomposition of the human body, as enjoined by the Buddha. The combination of modernity and the wisdom of the past, and the lack of any apparent disjunction between them – as seamless as the boy's robe – encapsulated the appeal of Santi Asok to urban, educated Buddhists.

Buddhadāsa's popularity has always been primarily among the urban, educated classes, though his this-worldly interpretations of cardinal Buddhist doctrines have paved the way for rural community development programmes as well. Phra Payom Kallayano is currently the best known of Buddhadāsa's disciples, and his supporters include a substantial number of young Thai. Payom was the first monk to show concern for people with AIDS in the late

1980s by visiting the Bumras Naradul Hospital in Nonthaburi where the initial sufferers were treated.

On Wednesday 17 November 1999, an M61 hand grenade exploded at the Wat Chanasongkhran in Bangkok. Police investigating the explosion linked it to the followers of Phra Dhammachayo, abbot of the Wat Phra Thammakaay, to the north of the city. Nobody was injured, two cars were damaged, and a puppy was killed by the blast. The incident was the culmination of a protracted struggle between the Wat Phra Thammakaay movement, whose members include large numbers of middle-class Thai, and the government, arising from allegations that the abbot had embezzled large sums of money, made false claims about miracles, and had stated that *nirvāṇa* is a physical location. The explosion was similar to a bomb attack on the house of Sathienpong Wanapok, an ex-monk and graduate of the University of Cambridge (UK), who had published criticisms of the movement.[31]

The beliefs of the Wat Phra Thammakaay movement are based on the teaching and meditation techniques of a former abbot of the Wat Paknam Phasi Charoen in Thonburi (now a major centre for *mae chii*). It follows the Mahāyāna concept of Three Bodies (*trikāya*), according to which the absolute nature of the Buddha is associated with his body of essence or truth, which is defined as *dharma-kāya* in Sanskrit (the basic language of Mahāyāna Buddhism), or *thammakaay* in Thai – hence the movement's name.

Thammakaay adherents part company with conventional Mahāyāna Buddhology by maintaining that the Buddha's body of essence is a spiritual entity which forms the basis of our physical and mental existence and is located in the region of the stomach. Meditation (followers refer to *samādhi*, rather than the more exacting *vipassanā*) reveals bright jewels which can glow so strongly as to be visible to others. Thus the abbot of the Wat Phra Thammakaay is often depicted as a celestial being with bright lights shining from inside his body.

A combination of exotic teaching and the absence of any critique of worldly comforts has made the movement very attractive to large numbers of well-to-do urban Thai. One university student described her impressions as follows:

> All of this was a real life experience that I will never forget. The . . . training was a progress to the beginning of a clear, new path, a new life for me . . . We were like people who had been born again, who had found a bright and secure life and a firm destination and goal.[32]

As in the West, such aspirations are more conducive to self-indulgence than to social and environmental engagement.

The essentially urban-based movements around Payom, Buddhadāsa, the Wat Phra Thammakaay, Kittiwuddho, the Engaged Buddhists (Sulak Sivaraksa) and Santi Asok were the subject of a questionnaire and interview investigation conducted among monks at Mahachulalongkorn Buddhist University in 1993. At the time of this investigation the monks would probably have had little knowledge of Samnak Paw Sawan or Hooppha Sawan, which are closely linked and were founded by Suchart Kosolkitiwong in 1973. During the last two years, however, under the guidance of Phichai Tovivich, a retired professor of chemistry at Chulalongkorn University, the Office of the World Peace Envoy, as the movement now calls its headquarters, has acquired a strong ecumenical and environmental thrust. Phichai has published articles on the toxicity of food additives such as monosodium glutamate, and encourages campaigns to oppose deforestation and prevent flooding.[33] The movement has set up a religious park in the Kaeng Krachan district of Petchburi province.

The culture of gender

Thai women play a large part in the activities of non-governmental organizations (NGOs) dealing with social and environmental issues, and support the *saṅgha*'s involvement in them. As yet, however, women are not permitted to ordain as full *bhikkhunī*, and the only comparable avenue open to them is the role of *mae chii*, which we translate as 'lay nun'.

It is claimed that Thai women have the highest level of work participation in the world.[34] Traditionally, women have always worked with men in the fields in addition to maintaining the household. Educational opportunities for women were initially the preserve of royalty and the upper classes, and during the second half of the nineteenth century boys and girls from wealthy families received identical primary education. The first students at Chulalongkorn University, founded in 1927, were seven women studying the arts and science. Most faculties currently have a majority of women.

Forty-five per cent of Thai women are employed in poorly paid agricultural or industrial work, which is the highest level in Asia. In the 44 to 49 age group, the percentage rises to 87 per cent, which means that the vast majority of Thai women continue to provide financial support for their families well beyond childbearing.[35] Different types of work are preferred by women in different regions. Thus in the central region they tend to run small businesses and do service work; in the north and south, which are relatively affluent, they work in the fields; and in the poorest northeastern region they participate in cottage industries (e.g. weaving) and domestic service. Many migrate to the capital.

Women therefore bear a heavy economic responsibility and also manage the family budget, but this does not necessarily lead to their having economic control, because, as Chatsumarn Kabilsingh points out, 'women may hold the family "purse strings", but in most cases, the purse is empty'.[36]

The subordination of women in Thai society is reflected in the legal system. Historically, this was largely shaped by the brahmanical repression of women as represented, for example, in the Laws of Manu. These state that a husband legally owns his wife and can sell her at whim. According to Thai Lanna law, the material value of a girl was exactly half that of a boy. These unjust laws were gradually reformed, and in 1974 equal rights for men and women were affirmed in the constitution, though not in law. Thai women were able to vote earlier than in some European societies (e.g. Switzerland), and the first woman member of parliament was elected in 1949. In 1981 there were fifty-four women members of the Thai parliament in a house of 301 – a higher level of representation than in Britain.

Chatsumarn Kabilsingh maintains that Indian and Chinese influences have largely shaped negative perceptions of women in Thai society. We have already noted brahmanical attitudes which, incidentally, are much more repressive than those of earlier Vedism in India. There were also additional prevailing Indian cultural values, such as the view that women must always be under the tutelage of a man: their fathers, husbands and even sons. Extensive trade relations with China developed during King Rama II's reign (1809–24), when many Chinese immigrants married Thai, and their cultural values were integrated with Thai beliefs. Kabilsingh maintains that traditional Chinese thought assigns five negative characteristics to women: they are easily manipulated, always unsatisfied, jealous, insulting and of lesser intelligence.[37] Some scholars of Chinese culture may wish to question her assessment!

Opinions differ over the extent to which Buddhist values are responsible for the low status accorded to women in Buddhist societies. Thomas Kirsch suggests that the Theravādin outlook constrains women to be more worldly and more prone than men to the desires that limit the realization of salvation.[38] Khin Thitsa disagrees with Kirsch's overall view, but agrees with him that Buddhism devalues women and serves as an ideology of oppression.[39] Charles Keyes disagrees with both Kirsch and Thitsa, neither of whom, he maintains, has 'probed deeply enough into the culture of gender in Thailand'.[40] The role that Buddhism plays in relation to female images is difficult to disaggregate from the total Thai cultural context. We have noted Kabilsingh's view that Thai Buddhism has been distorted over the ages by the gender biases of earlier Indian and Chinese societies. She also attributes the androcentrism of many Buddhist texts to the fact that they were recorded by men for men to read long after the Buddha's departure.[41]

Kabilsingh's claim that there is a 'pure Buddhism' which is gender-neutral is difficult to substantiate. The Buddha's essential teaching acknowledges that women are as capable as men of achieving enlightenment, but he seems to have been reluctant to accept women among his followers on an equal basis with monks, and when he did the eight additional rules (*gurudharma*) which he imposed on them were discriminatory.[42] However, we can no longer accept the Pali canon as containing the actual words of the Buddha, and must pay careful attention to parallel Mahāyāna texts, as well as particular sources such as the Therigatha, collections of early poems by Buddhist women, which Susan Murcott describes as testifying to a quality of spirituality 'based on the equality of men and women in the realm of the spirit'.[43] (These are included in the Pali canon.)

Serious practitioners of Buddhism may be *bhikkhu* (monks), *upasaka* (laymen), *bhikkhunī* (ordained women), or *upasika* (lay women). In Thailand, where the order of *bhikkhunī* has died out, there is a fifth category: women who wear a white garment, shave their heads and eyebrows, follow a form of monastic life and observe either five or eight precepts, called *mae chii*.

The term *mae chii* combines *mae* (mother) and *chii*, which refers to a white-robed ascetic (as distinct from a yellow-robed monk). Kabilsingh offers a variety of explanations for the precise meaning and origin of the term.[44] 'Nun' in English is not a good translation, because *mae chii* do not take comparable vows. 'Female ascetic', on the other hand, does not capture the monastic aspect of *mae chii* life. 'Lay nun' is probably the nearest English equivalent.

During the Buddha's lifetime there were both fully-ordained women (*bhikkhunī*) and women who led lives very much like *mae chii*. In Thailand there is no evidence for the existence of *bhikkhunī* during the reign of King Ramkhamhaeng (1283–1317), which probably means that the order had died out by then. In an unpublished thesis, Samer Boonma lists some of the possible reasons for their decline: an excessively long novitiate, brahmanical hostility and lack of political patronage.[45] The *bhikkhunī* order has never been revived, but there is evidence for the reappearance of *mae chii* from the seventeenth century onwards.

There is no precedent for *mae chii* initiation in the Tripiṭaka, but the requirements for membership are fairly similar to those for a monk. An aspirant must be a woman, must not be or become pregnant, must exhibit good behaviour, enjoy good health, be free from debt, be free from habit-forming drugs, must not be absconding from home or a government job, must not have a criminal record, suffer from infectious disease, be too old to perform religious duties, be lame, and must have permission to become a *mae chii* from her parents or husband.[46]

These requirements have been standardized by the Nun Institute of Thailand, which has its headquarters at the Wat Bovorniwes in Bangkok. The institute also regulates the rules for initiation into membership, which we may describe as ordination, though, strictly speaking, *mae chii* are lay women. A woman who fulfils the qualifications for 'ordination' goes to a *wat* and makes her request to the abbot. If this is granted, she will be put under the care of an abbot or senior *mae chii*.

The *mae chii* ceremony itself is conducted by four monks and several *mae chii*, and thus imitates full *bhikkhunī* ordination, since both male and female orders are involved. During the ceremony the aspirant is told that meditation is her highest religious duty and is reminded of the Three Refuges of Buddhism: the Buddha, the *dhamma* and the *saṅgha*. She is also given eight precepts; these are the five followed by all lay Buddhists prohibiting harm to any living being, stealing, sexual misconduct, lies and insults, and the taking

Role	Appropriate (%)
Teach at orphanage	74
Organize leadership training	65
Visit women prisoners	56
Teach in secondary school	55
Organize drug prevention	53
Give TV sermons	50
Operate AIDS helpline	39
Administer hospital	33
Teach monks meditation	29
Teach novices Buddhism	28
Work as midwife	9
Organize *mae chii* lottery	5

Figure 6.2 The appropriateness of social roles for *mae chii*

of intoxicants – which cloud the mind and hence inhibit meditation – plus three more. These additional precepts are to abstain from untimely eating (which means having the last meal of the day at noon), to abstain from dancing, singing, music, garlands, scents and all kinds of embellishments, and to avoid sleeping on a high or luxurious bed.

After ordination *mae chii* reside in monastic communities attached to *wats*, each having its own head *mae chii*. These communities may be quite small, consisting of half a dozen members, and are called institutes by English-speaking *mae chii*. Other communities, such as the one at the Wat Paknam Phasi Charoen in Thonburi, may contain up to 300 *mae chii* at any time.

Although meditation is officially acknowledged as the most important activity for *mae chii* – and some excel at it – many use their time to improve their educational qualifications and learn social service skills. In 1993 I conducted an investigation into the emerging social and pastoral roles of *mae chii* at three centres. One hundred and ninety-six *mae chii* completed questionnaires and thirty-five interviews were conducted. Figure 6.2 lists some of the social roles that *mae chii* regard as appropriate, according to Buddhism and society's needs.[47]

Mae chii are not subject to the Pāṭimokkha rules and can therefore assume roles which may be inappropriate for a monk. More than half stated that *mae chii* should teach children in orphanages, organize leadership training for lay women and men, visit women prisoners, and teach in a secondary school (some were concerned about their own qualifications for this). What is more surprising is that 53 per cent would be willing to organize drug prevention programmes, 50 per cent would give television sermons about Buddhism, 39 per cent would operate an AIDS helpline and 29 per cent would teach monks meditation! They were not asked specifically about the environment, though one 48-year-old *mae chii* from Chonburi province said that '*mae chii* should arrange meetings to help society to protect the environment, culture and tradition'.

Thirty-four per cent of the respondents believe that *mae chii* should be fully ordained as *bhikkhunī*, compared with 12 per cent of the young scholar monks at Mahachulalongkorn Buddhist University. Clearly there is a long way to go before Thai women can become full members of the *saṅgha*, but in the meantime many of them are playing an important if largely unrecognized catalytic role in Thai society as *mae chii*. As such they represent a vast, untapped resource with an enormous potential for resolving some of Thailand's major social and environmental problems.

We have considered southeast Asian and especially Thai resource depletion during the nineteenth and twentieth centuries, recognizing that although Thailand was never formally colonized, it none the less experienced a huge level of deforestation and the destruction of waterways. Since the early 1970s a small but vociferous 'green lobby' has protested against both the policies which have caused these problems and government attempts to respond to them with inappropriate 'development' such as large dams and eucalyptus plantations. The protesters have included local farmers, Buddhist monks and environmental NGOs.

Neighbouring southeast Asian countries such as Laos and Cambodia have experienced resource depletion comparable to that of Thailand, but their political histories during the past few decades render detailed evaluation difficult. Furthermore, insofar as our survey is of the relationships between religion and ecology, there seem to have been virtually no developments in Laos or Cambodia comparable to the activities of the Thai monks which I have described. The one exception is Phra Maha Ghosananda, a remarkable Cambodian monk who fled to Thailand following Pol Pot's excesses, and initiated some impressive community development programmes in the Thai holding centres along the Cambodian border in 1980.[48]

I summarized the historical development of Thai Buddhism against the background of its civic Ashokan legacy through various Siamese transformations to the nineteenth century, when the skills of Rama IV and Rama V prevented the European colonizing powers from gaining a foothold. Rama IV (Mongkut) played an important part in revitalizing Buddhism by reforming the *saṅgha* and attempting to provide the tradition with a rational and scientific basis, a task continued during the twentieth century by eminent lay scholars and monks such as Buddhadāsa.

From an ecological point of view, contemporary Thai Buddhism possesses many desirable features which are firmly based on tradition. These continuities with the past include the Buddha's central message, with its forthright rejection of the roots of human acquisitiveness, his affirmation of the value of non-human life, the frugality of monastic living, the ecologically based strictures of the Pāṭimokkha, and much else. Some traditional activities, such as the growing of medicinal plants in temple precincts, have acquired contemporary significance as people have come to realize the extent of global biodiversity loss. It is important for them to be encouraged and imitated elsewhere.

The transformations of tradition are more varied. As we saw in the last chapter, the Ashokan model of monarchy, with its emphasis on the cosmic role of the king, had considerable implications for the spread of Theravāda Buddhism in southeast Asia. King Mongkut's nineteenth-century reforms gave prominence to parts of scripture, such as the Pāṭimokkha, which stressed respect for all life and encouraged a more highly disciplined monasticism than previously. Although Buddhism had always been rational in the sense that it gave a logical analysis of human suffering and the way to overcome it, Mongkut provided it with a form of scientific rationalism which paved the way for sections of the *saṅgha* to rethink cardinal Buddhist doctrines.

I described a range of activities by monks, *mae chii*, and lay Buddhists, designed to overcome poverty and build self-reliant communities, and the manner in which they have increasingly begun to acquire an environmental dimension. *Mae chii*, whom some Thai still regard as little more than beggars, have begun to acquire new roles which are equipping them to meet a variety of social needs, illustrated in Figure 6.2. In fulfilling these roles they are able to complement and in some cases improve on the performance of monks, who are hampered by the restrictions of the Pāṭimokkha and the conservatism of the *saṅgha*. Modern Thai *mae chii* roles represent a major transformation of tradition, and are an enormous potential for social and environmental improvement. We also considered the activities of a number of urban Buddhist movements from the point of view of their social and environmental commitment.

A key factor in the transformation of Thai monasticism is the notion of appropriate behaviour, which I illustrated via the work of Phra Chamrun Panchan and his brother at the Wat Tham Krabok. Here a volatile community of young male and female drug addicts live in close proximity; the monks in charge of them are obliged to perform a variety of technically inappropriate functions, such as chasing after absconders. All this is acceptable, however, because the Buddha taught quite unequivocally that it is wrong to succumb to intoxicants and therefore, by implication, desirable to cure addicts.

The work of Phra Chamrun and Prawase Wasi represents the transformation of Buddhism within an essentially traditional framework. The same may be said of Phra Prajak and others who ordain trees and create sacred groves in order to frustrate the ambitions of commercial foresters, though Phra Dhammapiṭaka has taken issue with some of them. Few of these imaginative reinterpretations and transformations of Buddhism would have occurred, however, without the systematic influence of Buddhadāsa over several decades. In terms both of the lifestyles of ordinary Buddhists, lay and ordained, male and female, and at the level of scholastic and philosophical rigour, Buddhadāsa has provided Thai Buddhism with a credibility and a dynamism that is equipping it to address many of the most urgent social and environmental problems of modern Thai society.

7 India since Independence

During the opening of the new wing at St Stephen's Hospital in Delhi, the bishop was in the middle of a lengthy welcome to the chief guest, Prime Minister Indira Gandhi, when he noticed that she was looking at her watch.

'And we are so grateful that you have come to bless our Christian efforts . . .' His voice tailed off, as she tossed her sari round her shoulders and strode towards the microphone.

'You Christians do not need me to bless what you have been doing very well for a long time,' she snapped in the direction of the retreating bishop, proceeding to launch into a spirited exposition of the secular policies of her Congress Government.

India's secular state is distinctive because it is compatible with the promotion of religion by the established religious communities. There are limits, and the secular ideal has become increasingly tarnished, but the central government remains a bastion of the impartial governance of its subjects. However, in many respects, central government, however strong, simply cannot bring about social and environmental improvement for all. In this chapter I shall indicate the reasons for this and some of the problems which have accompanied India's attempts to transform itself into a modern nation state.

I shall indicate why successive governments, geared to a highly centralized bureaucracy, have been unable to surmount the social obstacles of caste, regionalism and the complexities of centre–state relationships. We shall study the ascendancy of the Hindu Right and its potential – as yet largely unrealized – for social and environmental improvement. Finally, we shall examine India's record in relation to international environmental politics and some of the proposals currently being made to resolve the most intractable social and environmental problems. Of these, we consider the views of Madhav Gadgil and Ramachandra Guha, as representative of an environment-led approach, and those of Jean Drèze and Amartya Sen as representative of a development-led initiative.

The secular state

Prior to Independence the Congress movement projected a secular ideology in order to present a broad and representative front to a society divided along the lines of caste and community. This was not, and is not today, seen as being incompatible with the assumption of Hindu symbols such as the notion of *Bharatvarsha*, the land of the Bhāratas, a Vedic tribe or people, or *Ram Rajya*, the kingdom of Ram, which featured in Gandhi's discourses. It did, however, have the unfortunate side effect of alienating the Muslim community (even before Independence), which resorted increasingly to an exclusive ideology.

Under the influence of Nehru, the Republic of India also acquired Buddhist symbols. A piece of sculpture known as the Sarnath Lion Capital facing visibly in three directions, and the wheel of *dharma*, both Buddhist representations, became emblems of the Republic. It seems likely that the chairman of the body set up to prepare the 1948 Draft Constitution, B. R. Ambedkar, who later became a Buddhist, was partly responsible for including these Ashokan legacies.

The Indian Constitution does not favour any particular religion and the head of state may be a member of any religious tradition or none. The Constitution promises 'to secure to all its citizens . . . liberty of thought, expression, belief, faith and worship'.[1] It also affirms that, as a sovereign, democratic republic, India will guarantee for all its members social and economic justice and equality of status and opportunity. The state may not make any distinction based on caste, creed or colour. Individual rights are guaranteed, but are balanced against national unity and security. While there is no suggestion of an equation between Buddhism and the secular state, there are points of affinity between the Ashokan state and the modern Republic of India.[2] However, 'secular' underwent a shift between Nehru, who understood it to mean neutrality towards religion, and Prime Minister Atal Behari Vajpayee, for whom it means the same attitude to all religions.

Strong centralized parliamentary government offered little scope for Gandhian notions of village republics, and the system of local self-government known as *panchayat raj* was largely glossed over in the Constitution.[3] There was no mention of socialism (initially, at least), none of the directive principles of the state could be enforced by law, and several of the fundamental rights of the Constitution could be overruled under special circumstances; thus property could be confiscated after the provision of compensation and citizens could be detained without trial for at least three months. Not only could the rights of individuals be curtailed, but entire states could lose their mandate to govern on the advice of a centrally appointed governor to the prime minister who, in consultation with the cabinet, could advise the president to declare a state of emergency and impose direct rule from Delhi.

Nehru's policy of strong rule from Delhi was underscored by three resounding general election victories: in 1952, 1957 and 1962. In order to win these elections, however, the Congress encouraged local alliances along caste, community and region rather than class lines. This made it difficult for underprivileged groups to mobilize effectively to gain access to state power.

The British had promoted the system of agricultural intermediaries known as *zamindari*. Under this system some land owned by *zamindars* was directly cultivated by landless sharecroppers (i.e. it was not rented by tenants). Other areas of land were maintained by tenant farmers who employed sharecroppers who were regularly paid but had no rights to the land. When the *zamindari* system was abolished following Independence, these tenant farmers became landowners.

Many of these landowning former tenant farmers belonged to what are officially known as the Other Backward Classes (OBCs), and support the Janata Dal political party. The landowning middle castes also include *jats* and *reddys*, i.e. the higher *śūdras* who are not twice-born but are not OBCs either. Significant numbers of these middle castes (both OBCs and non-OBCs) have gained economically through land reforms as well as by acquiring an education and diversifying their occupations. The lowest castes (i.e. untouchables, scheduled castes, *harijans*, *dalits* – there is no precise equivalent) have not gained as much as the middle castes, though they now have considerable political clout.

Nehru died in 1964, and by the time of the 1967 general election there was considerable opposition to Congress. The Jana Sangh, whose support came mainly from urban Hindu traders in the north, was opposed to Nehru's secular creed. The Praja Socialist Party, which broke away from a socialist group within Congress, polled the second highest number of votes in the 1957 elections, but failed to make inroads into the Congress's caste-based hold over the lower classes in rural areas in the north. The Communist Party of India had won the 1957 state elections in Kerala and had strong support in West Bengal and parts of Andhra Pradesh. In the early 1960s it split into pro-Moscow (CPI) and pro-Peking (CPI(M)) factions. Shortly after the 1967 elections the CPI(ML) emerged, with a revolutionary creed based on a combination of Marxism and Leninism.

The most important regional opposition to Congress came from the Dravida Munnetra Kazhagam (Dravidian Progressive Federation, or DMK), which emerged from pre-Independence anti-brahminism. That the Congress has been able to withstand sustained regional opposition from parties such as the DMK for so long is largely due to the strong relationship which has been built up between the Congress high command and the non-elected institutions of the state: the civil bureaucracy (e.g. the Indian Administrative Service or IAS), the police and the army.

Lal Bahadur Shastri, who became prime minister for two years in 1964, inherited a Congress leadership consisting of a motley and unrepresentative group of regional bosses and industrial capitalists, known collectively as the 'syndicate', who financed the party's election campaigns. Following his untimely death in 1966 the syndicate thought it had a malleable successor in Nehru's daughter, Indira Gandhi. In that supposition they could hardly have been more wrong.

'Abolish poverty'

Mismanagement by the party bosses, coupled with linguistic opposition to Hindi, were among the main reasons for Congress's crushing electoral reverses in 1967. The party retained an overall majority at the centre, but was swept out of power in eight states.

Indira Gandhi responded to the challenges facing her by following the policies of Nehru and Shastri of relying heavily on the civil bureaucracy, but she also attempted to strengthen the social bases of support for Congress through what came to be widely known as 'populist politics'. Before these strategies could be put to the test at the 1971 elections, however, she had to defeat her own party bosses. This she did in 1969 by a series of adept manoeuvres which split the party into two factions, forcing her to seek alliances with leftist and regional parties such as the CPI and the DMK to form a government.

Mrs Gandhi's new political alliances at the centre were accompanied by attempts in some states to mobilize support among the lower and more socially disadvantaged sections of society. In Gujarat, for example, she undercut one of her most resolute syndicate opponents, Morarji Desai, by appealing directly to *harijans*, *adivasis* and Muslims. She appointed a *harijan*, Jagjivan Ram, as Congress president, abolished the privy purses of the princely families and nationalized the banks, in an attempt to check the accumulation of wealth by the more powerful industrialists. Advisory committees were set up within Congress to monitor the problems of minorities such as the scheduled castes.

The slogan '*garibi hatao*' ('abolish poverty') was used with good effect during the 1971 general election, which was called, somewhat unconventionally, in advance of the 1972 state elections. Mrs Gandhi's Congress won a two-thirds majority in the Lok Sabha, and the state elections, coming in the wake of India's military intervention in East Pakistan, also reaped dividends. However, electoral rhetoric is never so easy to translate into action, and the former Congress bosses were able to mobilize their landed agricultural supporters against the centre. The judiciary challenged the legitimacy of bank nationalization, and in 1975 an unanticipated ruling from the Allahabad High Court invalidated Mrs Gandhi's own election result on the grounds that government machinery had been wrongly used in her support. The 1973 international oil price hike came at a particularly bad time for a government committed to stable food prices and high levels of employment.

In 1974 Jayaprakash Narayan, a much respected Gandhian socialist, assumed leadership of widespread opposition to Mrs Gandhi, who for her part resorted to more and more authoritarian measures, culminating in the formal declaration of an Emergency the following year. Among the worst excesses of the Emergency were Sanjay Gandhi's forced vasectomies in the name of family planning, and ruthless slum clearance in order to 'beautify' Delhi. Critics claimed that having failed to abolish poverty, Congress had decided to abolish the poor.

The 1977 general election swept Mrs Gandhi out of office and she reluctantly accepted the nation's verdict on her high-handedness. The coalition Janata Party which assumed power consisted of the former Congress Party led by Morarji Desai (who became prime minister), the Bharatiya Lok Dal, led by Charan Singh (a *jat* leader), the Socialists and the Jana Sangh (led by Atal Behari Vajpayee, more recently the leader of the Bharatiya Janata Party (BJP) and prime minister). J. P. Narayan, like Gandhi before him, held no office but wielded considerable moral authority. The Janata won the traditional Congress strongholds in the northern Hindi belt, while Mrs Gandhi's Congress (I) – 'I' for Indira – held the south.

The Janata coalition fell apart in 1979 and Morarji Desai was briefly replaced as prime minister by Charan Singh, who was unable to stem the tide of disenchantment. Mrs Gandhi once more made her broad-based appeal to the *harijans* and Muslims who had previously helped her to defeat Morarji Desai (though this time without the anti-poverty rhetoric) and won the 1980 general election by recapturing the northern Hindi states while holding on to the south.

Although Mrs Gandhi had successfully appealed to her widened social base, however, she did little to satisfy its members' material needs, and from 1982 until her untimely death two years later, she began to invoke policies of market-oriented liberalization. Under Rajiv Gandhi, Congress did extremely well in the 1984 general election, though not so well in the 1985 state elections – a negative consequence of the policy of unlinking state and national elections by holding the latter first, in the hope of putting pressure from the centre on the former.

Rajiv Gandhi avoided the anti-poverty populism and secular stance of his mother, thereby losing ground to the Janata Dal, which was concerned about justice for the agricultural backward castes, and the BJP, the former Jana Sangh in a new guise. He alienated both Hindu and Muslim communalists by introducing a constitutional amendment which nullified a Supreme Court judgement giving divorced Muslim women the right to claim alimony (the Shah Bano case). Muslims objected on the grounds that their cultural rights were being infringed (though 200 liberal Muslim women chained themselves to parliament in protest), and the Hindu revivalists denounced the Bill as an encouragement to Muslims to put religion before national allegiance. He further antagonized the Muslims by permitting Hindu communalists to lay the foundation stone of a Hindu temple on the exact site of the Babri *masjid*, an historic mosque in Ayodhya.

The 1989 general election went disastrously against Congress. V. P. Singh, who had deserted Congress towards the end of 1988 to form the Janata Dal, led an unstable coalition which combined communists on the left and the BJP on the right. In an attempt to consolidate his position he announced in August 1990 that he intended to implement the recommendations of the Mandal Commission. This meant that for certain central government posts, in addition to the existing reservation of 22.5 per cent for scheduled castes and tribes, an additional 27 per cent would be added for OBCs. The announcement provoked violence among upper-caste youth in many northern cities and there were several incidents of self-immolation.

V. P. Singh's coalition collapsed and in November 1990 Chandrasekhar was sworn in as prime minister of a minority government supported by Congress. This collapsed in March 1991. Rajiv Gandhi was assassinated by a member of the Liberation Tigers of Tamil Eelam in May 1991, following which Congress received a substantial sympathy vote in the elections and P. V. Narasimha Rao became prime minister, remaining in office until mid-1996.

The Congress under Narasimha Rao compromised its secular credentials by engaging in discussions with Hindu extremists, and although the Supreme Court had earlier issued a directive to the Uttar Pradesh government against demolition of the Babri mosque, it was destroyed in December 1992. According to Ayesha Jalal:

> The culpability of the premier nationalist party in the destruction of the mosque in December 1992, and the subsequent decision to build a temple to Ram amid the ruins, revealed in a glaring flash the full extent of the Indian state's structural atrophy and ideological bankruptcy.[4]

The period from Independence until the 1990s saw some significant improvements in the lot of ordinary people. The Congress forged local alliances along caste and regional lines which prevented the poorest sections of society from mobilizing effectively. The abolition of the *zamindari* system mainly benefited upper- and backward-caste landowners, who have so far successfully resisted attempts to further redistribute land. Indira Gandhi strengthened the social bases of support for Congress, but did little to check the power of the rich, with the result that her overall strategy played the top and bottom sections of society off against middle-income backward classes. V. P. Singh's reservations of government jobs for OBCs, who were already able to dominate the legislatures because of their numbers, was a tactic to win caste-based support and did nothing to empower the poorest social groups.

Rajiv Gandhi and Narasimha Rao moved the country in the direction of economic liberalization and deregulation in order to gain access to world markets, but there is considerable evidence to suggest that the benefits of this new emphasis have been unequally shared.

Patterns of resource use

The slogan '*garibi hatao*' ('abolish poverty') drew attention to economic deprivation but failed to acknowledge that poverty is only one of a number of fundamental deprivations. Development is a wide-ranging concept which Drèze and Sen describe as 'the expansion of real freedoms that the citizens enjoy to pursue the objectives they have reason to value . . . the expansion of human capability'.[5]

This view of development is not new and political economists such as John Stuart Mill would have had no difficulty with it. Nor would Nehru, when, on the eve of Independence in 1947, he urged the people of India to struggle for 'the ending of poverty and ignorance and disease and inequality of opportunity'.[6] During the years between the Second World War and the present day the notion of development has become dominated by overemphasis on growth in real income. It has also been naïvely assumed that growth in real income per head is proportional to the national income divided by the total population. We now know that national wealth does not 'trickle down' equitably to the poor.

If we understand development in its total sense, then India has been able to make considerable progress on many fronts since Independence. Famines – the last of which killed between two and three million people in Bengal in 1942 – have been eliminated (which is more than can be said for China, which in 1960 suffered a major famine in which thirty million people died). The multi-party system of democratic government has worked tolerably well and was able to withstand the excesses of Mrs Gandhi's Emergency in 1975. The creation of a large and capable scientific community has been accompanied by remarkable achievements in nuclear research and technology and the implementation of an impressive, if expensive, nuclear power programme. There have also been some unanticipated success stories, such as the 'White Revolution', whereby India had become in 1997 the world's biggest producer of milk.

Although these successes have been important, however, and have benefited large numbers of people, there are even larger numbers for whom virtually nothing has changed. The reasons are in part political – as we have seen – but are also the result of economic policies which have depleted non-renewable natural resources and produced unacceptable levels of pollution.

The colonial administration regarded India as a source of cheap natural resources and as a market for its own expensive commercial goods. Nehru's emphasis on industrialization was based on the belief that India must be as strong as her former oppressors. He was therefore not primarily concerned with the agrarian sector, which in any case was controlled by the states. Although agriculture was one of three main areas in which aggregate levels of public investment were promoted (the others being industry and infrastructure), this was because raw materials such as water and timber were needed as rapidly and cheaply as possible to build up basic and heavy industry.

India since Independence 117

The first three five-year plans (1950–5, 1955–60, and 1960–5) reflect these policies.

It was not until the fourth five-year plan (1969–74), well into Mrs Gandhi's premiership, that agriculture began to lead the economy. The plan envisaged the widespread use of technology to boost agricultural production which, given the nature of the patterns of land ownership, was bound to benefit the middle- to rich-income range of farmers. In 1968 US President Lyndon B. Johnson refused to deliver badly needed shipments of grain to India unless US agro-based industries were allowed to promote the sale of fertilizers and high-yielding grain varieties to these same farmers. The Bhopal disaster of 1984 was a consequence of the push for the excessive use of fertilizers and pesticides.[7]

Following Independence, state power was assumed by the new leaders, whose model of development is described by Gadgil and Guha:

> State power now passed into the hands of the landowner warrior and priestly castes of the countryside and the priestly and trader castes of the cities. This alliance was committed to the ideal of halting the drain of India's resources abroad. At the same time it was eager to refashion the pattern of resource use to serve its own interests. There were obvious limitations to what could be achieved in the framework of a low-input agrarian economy, one no longer capable of yielding much of a surplus. The solution obviously lay in industrialization; in tapping the energy of coal and petroleum, of hydroelectric power, in producing steel and cement and using the resources so generated to promote manufacture.[8]

Thus Nehru's vision of a new India dovetailed neatly into the less altruistic ambitions of urban and rural élites (which may not have been socially stratified as consistently as Gadgil and Guha maintain).

A similar alliance of convenience was to take shape in the agricultural sector:

> The way forward also lay in the intensification of agriculture, by irrigating large tracts of land under river valley projects and supplying them with synthetic fertilizer and pesticides . . . This was the model of development India opted for, rejecting the alternative, once offered by Mahatma Gandhi and some of his followers, of crafting an agrarian society of village republics making low levels of demand on the resources of the earth by living close to subsistence.[9]

Such a stark contrast between the Nehruvian and Gandhian paths to development does not do justice to the possibility of intermediate options. Otherwise, however, these authors provide a convincing account of the flows of materials and energy within several major ecological regimes (the sea, inland waters, forests, grazing lands and farm lands), and between these and the urban centres and abroad. In each of these ecosystems the overall effect is the same: natural resources are depleted (usually irreversibly) and the quality of life of local communities is diminished.

Gadgil and Guha summarize their arguments at the macro level as set out in Figure 7.1, p. 119.[10] This indicates that an alliance between three powerful interest groups has forced the country into a pattern of exhaustive resource use at the expense of the environment and the quality of life of the majority of ordinary people.

We shall consider briefly one of the major ecological regimes: forests. The colonial state declared vast tracts of forest reserved because it required timber to build ships and railway tracks and to service two world wars. People who depended on the forests for their livelihood were increasingly marginalized and denied their traditional community rights.

This fundamental pattern has changed very little since Independence. With reference to Figure 7.1, both the resource capital and the subsistence sector have continued to be depleted to satisfy the demands of industry (in India and abroad) and the urban population. The consumption of paper, plywood and polyfibre materials and the fuelwood needs of a steadily increasing population have increased the pressure on forests even more. In an attempt to meet such burgeoning demands new schemes have been adopted, some of which have gone badly wrong, as the following example illustrates:

> In the late 1950s foresters advocated giving up selective fellings in stands of natural forest, replacing them instead by aggressive plantation forestry, which called for large-scale clear-felling of natural forests to create plantations of fast growing exotics such as eucalyptus and tropical pine. This approach was adopted in the absence of any careful trials . . . Large tracts of prime rainforest, even primeval sacred groves, of the Western Ghats were cut down, only for it to be discovered that in these high rainfall areas eucalyptus falls prey to a fungal disease agent and may either die off or grow very slowly.[11]

The National Forest Policy of 1988 introduced some changes in the way forests were managed and took seriously the needs of forest dwellers, but the basic pattern of resource use is still the same.

Environmental politics

Nobody at the 1992 UN Earth Summit reading the Indian Government's official report, with its lyrical references to the Upanishads, sacred groves and Ashoka's pillar edicts, would have guessed how much political controversy had been generated during the previous two decades.[12] The same might also be said in anticipation of the period following the 1972 UN Conference on the Human Environment, at which Mrs Gandhi, almost as evocative in her

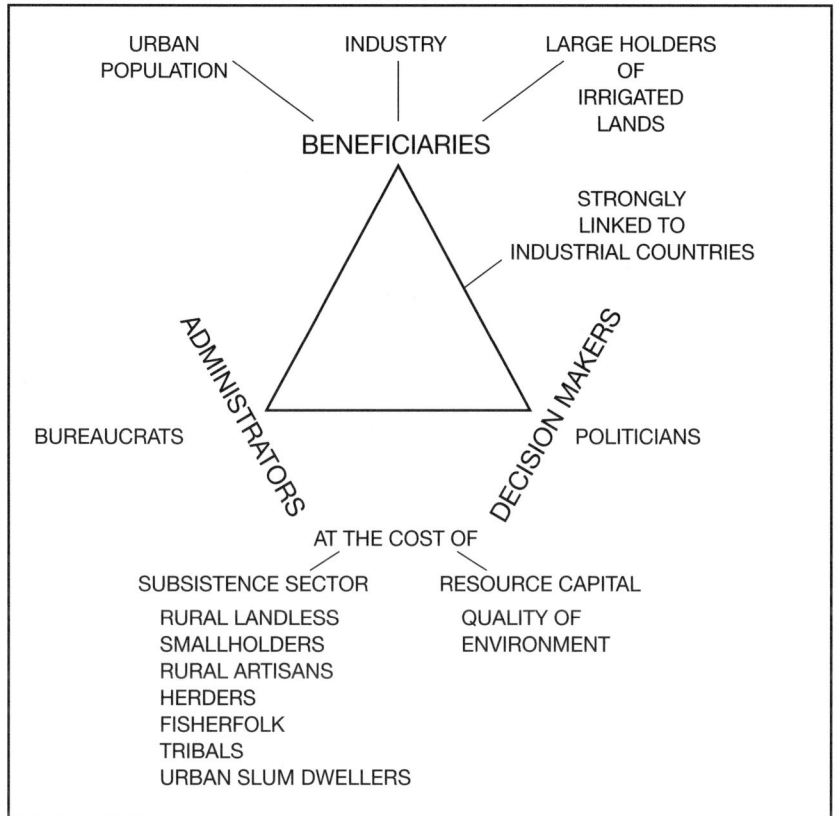

Figure 7.1 Patterns of resource use in India

Hindu imagery, presented a cogent account of India's needs. In this section I shall summarize India's contribution to international environmental deliberations between these two events.

Indira Gandhi was the only head of state or prime minister from outside Sweden to attend the 1972 UN Conference in Stockholm. She was one of the very few internationally known women political leaders and was riding on the crest of a wave of popularity for trouncing both her domestic enemies in the Congress syndicate and the Pakistani army in what became Bangladesh.

At the conference Mrs Gandhi expressed the view that:

> The environmental problems of developing countries are not the side effects of excessive industrialization but reflect the inadequacy of development ... [But] how can we speak to those who live in villages and in slums about keeping the oceans, the rivers and the air clean when their

own lives are contaminated at the source? The environment cannot be improved in conditions of poverty.[13]

Thus development, correctly understood, is basic to solving the problems of both poverty and environmental destruction; the two cannot be separated. However, the international community, dominated as it is by the most ruthless exploiters of natural resources, must assist the developing world in its attempts to protect the environment and alleviate poverty:

> Will the growing awareness of 'one earth' and 'one environment' guide us to the concept of 'one humanity'? Will there be a more equitable sharing of environmental costs and greater interest in the accelerated progress of the less developed world?[14]

Rejecting the simplistic view that high rates of population increase in developing countries are to blame for poverty and environmental degradation, she drew attention to the lifestyles and consumerism of the rich:

> It is an over-simplification to blame all the world's problems on increasing population. Countries with but a small fraction of the world population consume the bulk of the world's production of minerals, fossil fuels and so on. Thus we see that when it comes to the depletion of natural resources and environmental pollution, the increase of one inhabitant in an affluent country, at his level of living, is equivalent to an increase of many Asians, Africans or Latin Americans at their current material levels of living.[15]

Such sentiments chimed in conveniently with the '*garibi hatao*' slogans which had enabled Congress under Mrs Gandhi's leadership to win the 1971 general election. Elsewhere, both at the conference table and more informally, Mrs Gandhi and her colleagues reiterated their belief in national sovereignty and solidarity with the rest of the developing world.

Indian Government support for environmental issues declined following Mrs Gandhi's departure from office in 1977, but recovered three years later after she returned. On the recommendation of a high-profile committee headed by N. D. Tiwari (deputy chairperson of the Planning Commission), the Department of the Environment was set up in 1980 with the prime minister in charge. In 1981 Mrs Gandhi was one of a handful of heads of government to attend the UN Conference on New and Renewable Energy Sources in Nairobi. In 1985 the Ministry of Environment and Forests (MEF) was established.

The primacy given to development ensured that provision for the environment remained scanty throughout the 1980s. It received 0.04 per cent of the public sector outlay in the Sixth Plan (1980–5), rising to 0.4 per cent in the Eighth Plan (1992–7). Little was done to check the biggest industrial polluters:

petrochemicals, thermal power plants and coal mines, for example – all public sector activities. And whenever India's environmental interests were at stake at an international forum, the government was rarely represented by more than a member each from the MEF and the Ministry of External Affairs (which professes no environmental competence), plus members of the Indian Embassy of the host country.

Indian scientists were rarely consulted by the government on environmental matters and those who were tended to urge caution and further studies. By good fortune the only two NGOs which were taken seriously happened to be the most capable: the Tata Energy Research Institute, directed by Rajendra Pachauri, and the Centre for Science and the Environment, led by Anil Agarwal. In 1990 the latter convincingly challenged the accuracy of a report put out by the Washington-based World Resources Institute claiming that greenhouse gas emissions from developing countries were far more extensive than is the case.

In the years prior to the 1992 UN Earth Summit the Indian Government became involved in three major environmental issues: ozone depletion, climate change and biodiversity loss. The first was never of more than marginal importance to India, the second was not a major issue in itself but raised questions about the relative responsibilities between North and South, and the third concerned the fundamental problems of resource depletion discussed in the last section.

The first concrete evidence for ozone depletion over Antarctica was obtained in 1982. In 1985 the British Antarctic Survey published its results and claimed that the chlorofluorocarbons (CFCs) were largely responsible. Governments and multinationals in the West considered ways of reducing CFC emissions from various sources such as aerosol sprays and refrigerators, and in 1987 the Montreal Protocol set out targets for their phased elimination. The Protocol permitted developing countries to switch more gradually to feasible alternatives to CFCs and encouraged developed countries to provide them with technical assistance.

India did not take part in the run-up to the Montreal Protocol and sent only an observer to the proceedings, but after protracted negotiations over questions such as whether or not technology transfer could be guaranteed, India signed and ratified the Protocol in June 1992.

Climate change came to the forefront of international concern shortly before the World Conference on the Changing Atmosphere in Toronto in 1988. This time India was quicker off the mark, with the argument that since the industrialized countries were largely responsible for the problem, they should make the greatest efforts towards solving it. The final convention was signed by 154 countries, including India, at the Earth Summit. By this time India was in the throes of domestic economic upheavals and it was felt that it would be unwise to be too assertive.

The Convention on Biodiversity at the Earth Summit defines biodiversity as the variability among living organisms from all sources, including diversity within species, between species and of ecosystems. It was signed during the Summit by 156 countries, including India, but not the USA, though President Clinton agreed to sign it a year later after assuming office.

The conservation of biodiversity had become important in India several years prior to the UN discussions. The Wildlife Protection Act had been enacted in 1971 and subsequently amended twice, and Mrs Gandhi's personal support for Project Tiger in 1973 gave added prominence to the need to conserve wildlife. The Forest Act became law in 1980 and the Environment (Protection) Act in 1986. The related field of biotechnology was also of great interest to India because of the initial impetus it had given to the Green Revolution. In 1987 a government Department of Biotechnology was set up under the Ministry of Science and Technology.

During the 1970s developing countries began to question the system of common heritage which allowed northern plant breeders free access to the germplasm of developing countries. Matters came to a head in 1983 at a meeting organized by the Food and Agriculture Organisation. India wanted access to northern biotechnological information, but the North's understanding of intellectual property protection was incompatible with legislation whereby, for example, the 1970 Indian Patents Act recognized only process but not product patents. From the mid-1980s onwards there were conciliatory moves to reach agreement over the rights of plant breeders and patents, with the result that by the time of the Earth Summit only the USA remained dissatisfied.

There were gains and losses for developing countries, including India, prior to and during the Earth Summit. Their main problem was that from the outset the entire environmental agenda was set by the North, with the result that issues such as ozone depletion and greenhouse warming – which are only 'global' in the sense that the atmosphere happens to wrap itself round the earth – occupied pride of place, whereas much more pressing matters such as forest cover and the provision of clean drinking water were regarded as secondary or ignored. The conservation of biodiversity, however, which relates to crucial problems of natural resource depletion, was and continues to be a vital issue, and developing countries such as India were able to achieve significant gains in relation to it by utilizing the bargaining techniques they had used during the earlier debates. It is also significant that during the Earth Summit and in the run-up to it, India was able to collaborate closely with China, Pakistan and Malaysia, countries with which on more overtly political issues there have been serious differences.

The Hindu Right

We have earlier considered the Arya Samaj under the leadership of Dayananda Sarasvatī as a nineteenth-century reassertive religious response to secularization. Dayananda was a relentless critic of Islam and his movement appealed initially to lower middle-class Punjabis; it spread throughout the

Gangetic plain and beyond under the impetus of itinerant preachers supported by local bosses. Regional Arya *sabhas* advocated personal propriety (e.g. not spending too much on weddings), Vedic theism and cow protection, and tried to win over Muslims, low-caste Hindus and Christian converts.[16]

The Arya Samaj defined a new Hindu fundamentalism which paved the way for the Rashtriya Swayamsevak Sangh (RSS), founded by Veer D. Savarkar (1883–1966) in the mid-1920s. For Savarkar, an agnostic in private life, the Hindu religion was a small part of *hindutva*, or Hindu-ness, the unity of all Hindus. This unity was based on the late-nineteenth-century notion of *adhikari-bheda*, the differential rights and rituals of each level or group within the overarching Hindu hierarchy. However, Savarkar's definition of the Hindu in terms of a person's indigenous origin in a unique fatherland which is also the holy birthplace of religion minimizes distinctions between Hindu subgroups, which can therefore easily be mobilized together.

The RSS was given its name on Ram Navami Day, 1926. K. B. Hedgewar insisted on the term *rashtriya* (national) because he wanted to reassert the identity of Hindus within their geographical context, i.e. Hindustan. *Swayamsevak* was the name given to volunteers who attended the early training camps, and *sangh* is the collective membership.

V. D. Savarkar became the leader of the Hindu Mahasabha (great assembly), which was more concerned with providing a political alternative to Congress than with Hindu culture, which was the preserve of the RSS. Nathuram Godse, Gandhi's assassin, joined the RSS in 1930, deserting it a few years later for the Hindu Mahasabha. M. S. Golwalkar assumed leadership of the RSS in 1940, by which time the two organizations were largely independent of one another. By this time too, relations between Hindus and Muslims in north India were rapidly deteriorating.

The Vishwa Hindu Parishad (Universal Hindu Council or VHP) was founded in 1964 following a meeting between M. S. Golwalkar and a select group of Hindu leaders. Unlike the RSS, which claimed to represent the politically aware vanguard within Hindu society, the VHP asserts that it stands for the collective will of all Hindus, and since Hindus constituted almost 83 per cent of society at the 1991 census, it follows that the political will of the VHP must ultimately prevail. Dr Karan Singh was a founder member of the VHP.

As long as Congress dominated Indian politics the VHP was unable to exert much influence, but in 1984 – the year of Mrs Gandhi's death – an important gathering of members in Delhi called for the 'liberation' of three temple sites in north India. These were at Mathura, Varanasi (Benares) and Ayodhya, and it was decided to concentrate initially on Ayodhya, the birthplace of Ram in Uttar Pradesh, where a mosque was located. They would therefore have to move or destroy the mosque before constructing a temple to Ram.

The VHP's proposal to 'liberate' the Ayodhya site assumed that Ram had been physically born there and that a temple once stood at the exact location. It also assumed that in the sixteenth century the Mughal emperor Babar had destroyed the temple in order to construct a mosque. Many historians question these claims.

Prior to the call to liberate the Ayodhya site the VHP had been largely peripheral to national politics. In 1983 they had distributed the sacred water of the Ganges in three trucks all over India to symbolize national unity. In 1984 they issued a call for all Hindus to make bricks with the words 'Shri Rama' inscribed on them to be used to construct the new Ayodhya temple. At the same time they distributed huge numbers of stickers depicting Ram, the temple and the sacred syllable *OM*. Ram was usually portrayed either as a handsome warrior ready to die for his homeland, or as a cherubic child held prisoner in a Muslim building at his birthplace.

The Bharatiya Janata Party (Indian People's Party or BJP) grew out of the Jana Sangh, which had earlier been closely associated with the RSS, though Deen Dayal Upadhyaya had tried to give it a more flexible ideology which was less antagonistic to Muslims. Upadhyaya also put forward his view of integral humanism, which was opposed to both western capitalist individualism and Marxist socialism, though welcoming to western science.

At the end of Mrs Gandhi's Emergency in 1977 the Jana Sangh was able to obtain ninety-three parliamentary seats in the Janata coalition. In 1981 this had dropped to sixteen, and the newly constituted BJP, which replaced the Jana Sangh, only got two seats in the 1984 elections. Throughout this period the RSS continued to grow in numbers, however, and was able to extend its support into the four southern states.

The success of the brick project encouraged the VHP to undertake an even more ambitious procession to Ayodhya in 1990. L. K. Advani, leader of the BJP, rode on Rama's chariot (a converted Toyota) through eight states, from the Gujarat coast to Ayodhya. The *rath* (chariot) did not carry a consecrated image or any picture of Ram, but the procession or *Rath Yatra* quickly assumed a ceremonial and devotional character of its own. Moderate BJP politicians emphasized the moral and petitionary dimensions of the procession, comparing it to Gandhi's 1930 salt march. Advani distanced himself from the more volatile members of the Bajrang Dal, the youth wing of the VHP, by admonishing them not to carry tridents and other primitive weapons.

Between L. K. Advani's *Rath Yatra* towards the end of 1990 and the destruction of the Babri mosque on 6 December 1992, there were extensive riots in which large numbers of people, mostly Muslims, were killed. Advani was arrested in Bihar, whereupon the BJP withdrew its support from V. P. Singh's coalition, which collapsed, necessitating fresh elections. During the election campaign Rajiv Gandhi was assassinated, and the consequent wave of sympathy for Congress brought P. V. Narasimha Rao (jailed in October 2000 for corruption) into power. It was under his premiership that the mosque was destroyed. Throughout this period and subsequently the Hindu Right has become increasingly influential.

The BJP's electoral successes in the 1998 and 1999 general elections were to some extent the culmination of efforts to form a Hindu nation-state based on indigenous culture by members of the Sangh Parivar (Sangh family): the VHP, Bajrang Dal, the BJP and the RSS. Granted that the Sangh Parivar has devoted most of its energies to questionable political goals centred around a highly selective interpretation of the Hindu tradition, it is significant that there have been a few references to ecology by its leaders. Tapan Basu et al., the authors of a perceptive critique of the Hindu Right, cite one such example:

> In the self-image of the RSS today, it appears, [Golwalkar's] speech constitutes the 'philosophical' foundation of a Hindutva brought up to date to suit contemporary conditions. This is a 'positive' Hinduism, based on '*angangibhava*' ('limb-body' relations – roughly, organicity). A hierarchy is postulated of individual-society/family-nature-divinity . . . and men can reach God only through reverence for society and nature. Society is immediately identified with familiar relations, which would evidently provide the paternalist model for all other relationships, including that between rulers and ruled. Contrary to Western ideals, we are told, this is a philosophy which gives precedence to duty towards the community over individualism and materialism. Western individualism shatters family and community, Hinduism integrates them through a harmonious *dharma* – which . . . rests on the right balance between four *shaktis*: intelligence, power, wealth and labour.[17]

Tapan Basu and his co-authors contrast Golwalkar's foundation of *hindutva* with Deen Dayal Upadhyaya's integral humanism:

> In 1965, Deen Dayal Upadhyaya tried to give the Jana Sangh an ostensibly distinct ideology and impart a veneer of flexibility and openness to the Savarkar-Golwalkar framework through a series of lectures to party members on 'integral humanism'. Upadhyaya claimed to be 'scientific', welcomed Western science (as distinct from Western 'ways of life'), and declared that even the principles of *dharma* may have 'to be adapted to changing times and places'. Certain economic objectives – like full employment and free education and medical treatment – were mentioned for the first time, without specifying concrete methods for realizing such laudable goals. There was little use of the word 'Hindu', and no obvious abuse of Muslims . . . [However] change has to be in conformity with 'our culture that is our very nature' and here '*Bharatiya*' integral humanism is opposed to both capitalist individualism and Marxist socialism, for these are based on the harmful Western idea that progress comes through conflict. The ideal, in contrast, is one of harmonious relationships everywhere, as between a body and its limbs, applied to man and nature, individual and society, labour and capital: an obvious echo of

Golwalkar's *angangibhava*. In an interesting gloss on Savarkar, the 'underlying unity' of *'bharatiya* culture' is located, not so much in a place of origin as in a distinctive 'soul' or 'identity', and the 'laws that help manifest and maintain' its inner essence constitute the *'dharma'* of the nation.[18]

Ecology and the 'spoliation of nature' also find their place in Upadhyaya's thinking:

> In Upadhyaya's *Integral Humanism* there was a small, undeveloped reference to hedonistic consumption and the consequent depletion of natural resources. Interestingly, recent writings have shifted the focus to a more trans-social notion of a rather crude form of ecology, which traces the spoliation of nature to the innate character of Western scientific man and to his desire to conquer nature. The delinking of ecological problems from consumerist capitalism constitutes an absent argument which is illustrative of the relevance of new Hindutva for a diasporic, urban, business-oriented, booming global formation fuelled by consumerist desires and opportunities. The counter-order is located in a superior 'scientific' organic understanding within traditional Hinduism which structures man within, and not against nature. Hindutva today is in a self-congratulatory mood. While *Integral Humanism* did concede that Western science has something to teach us, more recent RSS-VHP publications are more at pains to establish that Western scientists themselves are rediscovering the relevance of Hindu social, physical and human sciences.[19]

K. S. Sudarshan, general-secretary of the RSS, claimed in an interview with Tapan Basu that 'true Hinduism is based on *srṣṭi-dharma*, which is nothing but ecology. The individual and society must seek a proper balance with *prakṛti.*'[20] However, Basu and his colleagues also point out that the RSS 'has so far displayed no active concern or interest about the ecological movements going on within the country'.[21]

Restructuring society

In this section we consider two programmes to restructure Indian society in order to reduce poverty and the abuse of nature. The first, by Madhav Gadgil and Ramachandra Guha, approaches these problems from the point of entry of environmental abuse.[22] The second, by Jean Drèze and Amartya Sen, is primarily concerned with economics, but it will be argued that if implemented on a large scale it would provide an appropriate framework for the resolution of most major environmental problems.[23] Both pairs of authors offer important contributions to the environment and development debate; the first makes 'environment' the point of entry, the second 'development'.

Gadgil and Guha draw attention to the contradictions between the ideal India of culture and tradition, and the harsh realities experienced by many people:

> On getting up in the morning, we are expected to beg forgiveness from Mother Earth for stepping on her . . .
>
> O earth, consort of Vishnu, the Lord of creations, with
> mountains for thy breasts, and oceans for thy garments,
> forgive me for stepping on you.
>
> But not only do we not mind stepping on the earth, we blithely tolerate disasters like the Bhopal gas leak. India gave birth to Gautama Buddha, in a sacred grove of sal trees dedicated to the goddess Lumbini; Buddha achieved enlightenment under a peepul tree and preached a doctrine of compassion towards all creatures on earth. Today we are cutting down peepul and banyan trees, protected over centuries, to bake bricks to build our cities and to crate mangoes sent to the Middle East. We respect Mahatma Gandhi as the Father of the Nation; above all he wanted independent India to be rejuvenated as a land of village republics. But over the past forty-eight years, we have systematically sabotaged attempts to empower village people to control and manage their own destiny.[24]

These authors attribute the system-wide difficulties to six root causes. These are:

1. The natural resource base on which most people depend is becoming increasingly circumscribed. For example, natural forests on which communities depend for their basic needs are giving way to commercially valuable eucalyptus or acacia plantations.
2. The majority of people have very little access to the human-made resources of the organized industry-services sector. This is because employment in this sector has grown more slowly than the population and because education has failed to equip people for skilled work.
3. The state monopoly has produced human-made capital inefficiently and without public accountability. Natural resources have been subsidized and the environmental costs of the destruction of natural capital have been passed on to the poor.
4. The rich, in collusion with state power, are establishing an even stronger hold over natural capital. For example, the refugees from the Narmada Dam are being displaced against their wishes and without appropriate plans for their resettlement.
5. People who are excluded from human-made capital and education have

128   *Religion and Ecology in India and Southeast Asia*

no incentive to invest in quality of offspring and therefore produce large numbers of them.

6   Heavy dependence on imports of technology and petroleum products and the consequent need to export in order to pay for them are producing large-scale outflows which adversely affect natural capital, e.g. iron and manganese mining silting up estuaries, overfishing in the sea and overgrazing to produce leather for export.

In summarizing these six points I have not followed Gadgil and Guha's division between 'ecosystem people' and 'ecological refugees' on the one hand, and 'omnivores' on the other, because it seems that people cannot be so clearly distinguished. For example, the urban middle classes in major cities may be omnivores from the point of view of their general standard of living, but when their children brave poisonous traffic fumes on buses each day to get to school, they are ecosystem people if not ecological refugees. I have therefore avoided this terminology altogether.

The authors proceed to consider the extent to which three particular political philosophies are comprehensive enough to address the six root environmental problems. These are the Gandhian, Marxist and liberal–capitalist schools of thought.

According to the Gandhians, environmental problems are caused primarily by materialistic greed. This is why the natural resource base is less than adequate for human needs (point 1). Greed is also the reason why many people have little access to human-made resources (point 2) and why the state does not fulfil its role satisfactorily (point 3). Gandhians believe that the rich should not increase their levels of consumption and must give up attempts to secure a stronger hold over natural and human-made capital (point 4). In a decentralized system of village republics people will not be excluded from natural and human-made resources, their own needs will be reduced, and they will be able to invest in quality of offspring rather than large numbers (point 5). Finally (point 6), India should halt the drain of natural resources to international markets by doing away with the need for foreign exchange to purchase technology and military hardware.

Marxists view environmental problems as the result of capitalist exploitation of both people and nature. They have no problem with the increased use of natural resources as such, provided they are utilized by the state to satisfy basic human needs. Science is to be encouraged as a means of promoting these ends.

With reference to Gadgil and Guha's six points, Marxists believe that ordinary people should be given far greater access to and control over the natural resource base (point 1). In relation to point 2, they should also have more access to human-made capital (e.g. basic industrial commodities). The production of human-made capital from natural resources should be accomplished under conditions of public scrutiny, efficiency and the equitable sharing of environmental costs (point 3). Marxists believe in limiting the powers of the rich (point 4), and in the equitable distribution of human-made capital (point 5). Finally, Marxists will reduce the drain of natural resources as a consequence of international capitalism (point 6).

The liberal–capitalist approach to environmental problems envisages their reduction through the use of clean and efficient technologies which promote the efficient conversion of natural resources into human-made capital. The world capitalist system, with its commitment to the expansion of trade and economic liberalization, is unlikely to improve the access of ordinary people in a developing country to their own resource base (point 1). According to capitalist economic theory the benefits of expanding markets should 'trickle down' to people who currently have very little access to the human-made resources of the industry–services sector (point 2). In practice this does not happen. The pruning of state bureaucracy and the removal of subsidies on natural resources have proved difficult to achieve (point 3). Private enterprises which take over from the state may pollute just as badly. Liberal–capitalism does not reduce the power of the rich (point 4), or concern itself about which people are doing skilled jobs, as long as there are enough people to do them (point 5), and it will further accelerate the drain of natural resources abroad (point 6).

Each of these three political philosophies has strengths and weaknesses. Gandhism relies on voluntary restraint, which is unlikely to appeal to the majority of people without a strong element of self-interest or compulsion. Marxism in Eastern Europe and the former Soviet Union has produced an immensely inefficient and environmentally wasteful state apparatus, and liberal–capitalism, though more efficient and open to scrutiny, is unlikely to lead to the level of empowerment that will enable ordinary people to look after their own and nature's interests:

> The philosophy of economic liberalization does have as its core three significant themes: namely, the need to do away with state-sponsored subsidies; to prune the power and size of the state apparatus; and to create an open, democratic society that should in theory imply a better deal for India's environment and people. Implemented in the context of a highly inequitable society, however, this philosophy is unlikely to lead to any genuine progress.[25]

Gadgil and Guha maintain that the six root causes of environmental degradation can be removed by measures based on desirable components of each of the three philosophies. They list nine such measures, which I summarize:

1   India must practise a form of participatory democracy in which the political leadership as well as the bureaucracy is accountable to ordinary people. This will involve strengthening *mandal panchayat*s (i.e. five-member councils for a group of villages) and *zilla parishat*s (i.e.

assemblies of elected members governing a district). These need to be protected from political interference.

2 Local populations should have a major voice in the process of control, planning and implementation of natural resource use. They should be encouraged to use their traditional knowledge. Hill people in Manipur have agreed among themselves not to harvest shoots or market bamboo from village forests because they need it for their own house construction.

3 Local communities should benefit from profits realized from the utilization of natural resources such as timber, granite, coal and oil, which must be valued at realistic prices. This will encourage environmentally sound methods of extraction and efficient resource use. A proper price tag must be attached to any unavoidable degradation of the environment, charged directly to the agency responsible, and the money should be ploughed back into the promotion of environmental awareness in schools and colleges.

4 Points 1 to 3 can only be achieved in a more equitable society. In a predominantly agricultural society radical land reform is essential.

5 The experience of Eastern Europe suggests that state socialism is wasteful. The shift towards private enterprise in India should be continued, state subsidies should be removed, and pricing mechanisms should be promoted that will encourage environmentally friendly policies.

6 The scale of economic enterprises, which is currently biased towards macro projects, should be geared to participative processes of decision-making. Small is not necessarily beautiful, but with the value of hindsight several of India's large dams could have been replaced by a larger number of intermediate hydro-electric projects.

7 Communication technology should be utilized to enhance the scope for people's participation in monitoring the environment. For example, satellite imagery can easily make available to ordinary citizens up-to-date information about forest cover or the silting of river beds. People's science movements such as the Kerala Shastra Sahithya Parishad should be encouraged. (This organization, which was founded in 1962 and currently has more than 62,000 members, took part in the Earth Summit.)

8 Human demands on the resources of the earth have escalated as a result of increases in population and per capita consumption. The transition to smaller families will occur when a more equitable development process draws more people into the ambit of modern industry, services and agriculture. The rich must accept restraints on their levels of consumption.

9 India's existing high-cost, mediocre quality economy (in certain sectors), with its reliance on imports of petroleum and military hardware, plus consequent borrowing and foreign debt, must be replaced by new strategies for agriculture and industry requiring lower levels of energy input and an improvement in international relations conducive to reduced expenditure on defence.

Gadgil and Guha summarize the pros and cons of the three contending political philosophies in relation to these nine desirable measures:

> Each of the three contending philosophies – Gandhism, Marxism and liberal capitalism – thus has components which are very desirable when viewed from an environmental perspective. But each philosophy also has components that deserve to be decisively rejected. What is evidently needed is a synthesis of the several positive elements: decentralization and empowerment of village communities along with a moderation of appetite for resource consumption from Gandhism; equity and empowerment of the weaker sections from Marxism; and an encouragement of private enterprise coupled to public accountability in an open, democratic system from liberal capitalism. In so far as Gandhism seeks to conserve all that is best in our traditions, it might be called the Indian variant of conservatism, with this significant caveat: that it seeks to conserve not the hierarchy of aristocratic privilege, but the repository of wisdom and meaning vested in ecosystem people. Marxism is of course the best-known strand within the socialist movement, while democratic capitalism is the ideology of the liberal. If our arguments are correct, then the environmental philosophy most appropriate for our times is, in fact, nothing but conservative-liberal-socialism.[26]

Beyond liberalization

Consistent with their view of development as the expansion of human capabilities, Drèze and Sen argue for a broad and participative interpretation of economic development which takes into account the need to expand social opportunities. They recognize that where social opportunities already exist, the enlargement of markets can significantly enhance them, but that most people are excluded from benefits through lack of literacy and education, together with other capabilities associated with basic health, social security, gender equality, land rights and local democracy:

> With nearly half the people – and close to two-thirds of the women – illiterate, the transformation of the Indian economy is no easy task ... The success of the east Asian 'tigers', and more recently of China, has been based on a much higher level of literacy and basic education than India has.[27]

Granted that India has made important gains in a number of areas – life expectancy has increased from about thirty years at Independence to sixty years today – there are others in which hardly any progress has been achieved, and this is especially true of elementary education. The reasons for India's successes and failures compared with other countries cannot be attributed to varying economic policies. Countries which are usually considered to have moved ahead of India economically have pursued quite different policies, ranging from market-oriented capitalism (South Korea, Taiwan, Thailand), to communist-style socialism (Cuba, Vietnam, China) and a variety of mixed-economy systems (Sri Lanka, Jamaica, Costa Rica).

It is not correct to attribute India's failures to increases in population: Drèze and Sen point out that if China had a population growth rate equal to India's 2.1 per cent instead of its own 1.4 per cent, the growth rate per capita would only decline from 7.7 per cent per year to 7.0 per cent (assuming no change in the growth rate of domestic product). Similarly, if India should succeed in bringing down its rate of population growth from 2.1 per cent to China's level of 1.7 per cent, the growth rate per capita of Gross Domestic Product would only increase from 3.1 per cent to 3.8 per cent. The authors point out that population pressure on the local environment and on the quality of life of women is more serious than its effects on economic growth.[28]

Drèze and Sen also draw significant conclusions from internal comparisons between different parts of India. Whereas the female literacy rate in Kerala is 86 per cent, in Rajasthan it is only 20 per cent. Taking into account gender and caste, comparisons are even worse. Thus, whereas 94 per cent of men in Kerala are literate, the literacy rate of scheduled caste women in Rajasthan is well below 10 per cent. Although southern states such as Kerala are generally better off than the northern ones, however, gender inequalities cannot be simply correlated against affluence, and are most acute in the northwest, which includes the relatively prosperous states of Punjab and Haryana.

The authors give five reasons why education and health are vital components of human freedom to develop:

1. Education and good health are intrinsically important and the opportunity to have them enhances a person's effective freedom.
2. Education and good health can facilitate many things such as getting a job.
3. Literacy and a good education can promote public discussion of social needs and encourage informed collective demands which can in turn expand public facilities.
4. The process of schooling can foster important auxiliary benefits such as the reduction of child labour and the broadening of social horizons, especially for girls.
5. Literacy and a good education on the part of disadvantaged groups can increase their ability to organize politically and resist oppression.[29]

Through these interconnections, education and health can be of enormous strategic importance for economic development, but compared with the high performing Asian economies in which the state has played a large part in promoting education, the Indian planners have been extremely dilatory.

Drèze and Sen comment on the large body of Indians with higher education, and recent successes in utilizing their talents in computer software and related industries. However, this makes little difference to the much larger numbers of poor, uneducated people:

> The abysmal inequalities in India's education system represent a real barrier against widely sharing the fruits of economic progress, in general, and of industrialization, in particular, in the way it has happened in economies like South Korea and China – economies which have succeeded in flooding the world market with goods, the making of which requires no great university training, but is helped by widespread basic education that enables people to follow precise guidelines and maintain standards of quality . . . It may be much less glamorous to make simple pocket knives and reliable alarm clocks than to design state-of-the-art computer programmes, but the former give the Chinese poor a source of income that the latter does not provide.[30]

These authors also draw attention to a tendency towards elitism in the Indian educational system which they attribute to the Hindu and Islamic traditions:

> Both Hinduism and Islam have, in different ways, had considerable inclination towards religious elitism, with reliance respectively on Brahmin priests and powerful Mullahs, and while there have been many protest movements against each (the medieval poet Kabir fought against both simultaneously), the elitist hold is quite strong in both these religions. This contrasts with the more egalitarian and populist traditions of, say, Buddhism. Indeed, most Buddhist countries have typically had much higher levels of basic literacy than societies dominated by Hinduism or Islam. Thailand, Sri Lanka and Myanmar (Burma) are good examples.[31]

The removal of ignorance, illiteracy, remediable poverty, preventable disease and inequalities in opportunities are objectives to be valued both for their own sake and as a means to economic growth:

> The crucial role of wide participation is central . . . and in so far as this requirement is missed out in policy discussions, that omission calls for rectification. At the same time, we must not confound that requirement with a rejection of the importance of economic growth itself. There has to be growth for it to be participatory.[32]

The authors cite Nagaland as a good example of participatory development:

> Tribal societies in India, which tend to be relatively egalitarian, have a long tradition of participatory local democracy. In some states, notably

Nagaland, this tradition has formed a good basis for participatory programmes of economic development and social change. The relatively cohesive nature of many tribal societies in India has also been conducive to diverse types of collective action, ranging from environmental protection to resistance against forced displacement.[33]

The Nagas are mainly Christian, and their egalitarianism may be related to their religious beliefs (though Meghalaya, which is also Christian, is not egalitarian). Christianity has been blamed for environmental destruction in the northeast on account of its tendency to desacralize trees, sacred groves, etc. Here, however, it can be credited with laying the foundations for community action to protect the environment. Drèze and Sen continue:

> We have paid particular attention to specific activities that demand priority attention at this stage, such as the expansion of basic education. But there are many other relevant fields of action, including not only those which we have had occasion to discuss (e.g. health care, social security, population policy, land reform, local democracy, women's rights), but also others . . . e.g. the need for sound environmental policies, for improved rural infrastructure, or for a credible legal system. The key issue is that many of the basic entitlements that people need in order to improve their lives (e.g. to primary education, child vaccination, safe contraception, clean water, social security, environmental resources, legal protection) depend to varying extents on some sort of positive government activity. The positive role of the state is, thus, potentially quite extensive.[34]

Drèze and Sen proceed to give examples of states such as West Bengal and Kerala where a combination of responsible government and public activism has brought about significant improvements. Within these improvements, mainly at local and district levels, there have been some important environmental initiatives by groups and individuals which I shall summarize in the next chapter.

We have considered some of the features of the post-Independence transformation of India into a modern nation-state, noting the interplay between central politics and regionalism, the national movement and the forces of communalism, and the manner in which a combination of inherited bureaucracy and industry-led planning perpetuated natural resource depletion and did little to improve the lot of the poor.

From the early 1970s onwards, under the firm direction of Indira Gandhi following her enthusiastic participation in the UN Stockholm conference on the environment, India became increasingly involved in global environmental politics and played an important part in the Earth Summit held in Rio de Janeiro in 1992. However, the UN agenda, which had been set by the western industrial nations, failed to address the fundamental problems of deforestation, lack of clean water supplies and natural resource depletion which are endemic in India as in many other developing countries. None of India's existing political parties appears willing to confront these issues directly.

We have also considered two sets of proposals for ameliorating these problems, the one by Gadgil and Guha which recommends an environment-led approach based on a combination of the political philosophies (including Gandhism) which emerged during the Chipko movement, the other by Drèze and Sen which advocates participative development at local and district levels concentrating on basic education, healthcare and a reduction of the gender imbalance. Within such a context empowered communities will protect their natural environment using whatever resources are available. Some of these will be modern and scientific (e.g. satellite imagery, as suggested by Gadgil and Guha), while others will be traditional (e.g. sacred groves, historic methods of river management).

My analysis of the socio-political and economic changes that have occurred in India since Independence does not lend itself to comparisons between tradition and modernity as in previous chapters. However, the use of Buddhist symbolism to represent the secular state is a powerful evocation of Ashoka's self-understanding as the universal guarantor of social and environmental equity. What Ashoka could accomplish by royal fiat, however, can only be accomplished by his modern successors via the longer and more difficult route of participatory democracy.

8 Signs of hope

According to Tavleen Singh, an influential columnist in *India Today*:

> Environmentalists are powerful creatures these days. They can stop dams from being built, halt highways in their tracks . . . block power plants before these can produce a single watt of electricity and, in fact, bring all development to a grinding halt. The environmentalists' efforts are hugely popular with the press . . . Most of these gentlemen – and ladies, not to forget Medha Patkar – belong to the educated middle classes who enjoy such twentieth-century amenities as electricity. Yet, try putting up a power plant and they descend in swarms, hackles bristling in the polluted air.[1]

Not to be outdone, Medha Patkar (who has recently been joined in protesting against the Narmada Dam by Booker Prize winner Arundhati Roy) hit back at her critic:

> Tavleen [Singh] asks whether or not the basic twenty-first-century amenities should reach every poor family. Whom does she address this question to? Those of us who are striving to take basic education, community health centres, minimum sources of energy and other such amenities to the ever-neglected tribals, the rural poor and women facing drudgery? . . . Tavleen herself is in the dark as far as this dark side of poor India is concerned. She represents a mindset which sees each and every typical development symbol – huge dams, power projects, urban concrete jungles – as naturally beneficial . . . This is typical of an urban energy-exploiter, who buries the truth of inequality – and even destruction of lives and livelihood of the majority – under the glittering life in his or her backyard.[2]

But Tavleen Singh remained adamant:

> What Patkar is ensuring is that the poor will continue to live without even basic twenty-first-century amenities . . . To cater to the needs of a billion

people by the first decade of the next century, we need an estimated 60,000 MW of additional electricity just so that every Indian can light bulbs and work a fan. If we do not succeed then it will be the poor who will suffer.[3]

Beneath the vitriol one suspects that Medha Patkar and her associates – though possibly not Arundhati Roy – have a better understanding of the basic needs of ordinary people and ways of satisfying them. In this chapter I shall discuss a variety of local initiatives, ranging from the indigenous NGOs which went to considerable lengths to transport themselves to Rio de Janeiro to participate in the Earth Summit, to people who are engaging in imaginative environmental activities in their home situations. It is my contention that the work of these activists offers signs of hope which, if taken up on a wider basis, will enable us all to look forward to a more sustainable future. I also indicate that religious and cultural traditions play a significant part in these initiatives.

I begin by summarizing the main characteristics of the Indian groups and organizations which participated in the official proceedings at the Earth Summit.

Global-level initiatives

The list in Appendix B gives brief details of the hundred Indian NGOs which were fully accredited to the Earth Summit. In each case the name of the organization is given, sometimes with a translation from the Hindi or other regional language (provided it is reasonably straightforward). Thus Yuva (No. 100) means 'youth', which is appended, whereas Taralabalu Rural Development Foundation (No. 90) translates less easily as 'star-like children's rural development foundation', which has not been added. The location of the headquarters of each NGO is given, together with the state, unless a city is very well known (e.g. Calcutta). Some major cities have recently changed their names: Bombay is now Mumbai and Madras is Chennai. There is a map of the Indian states at the end of Chapter 1 (Figure 1.1). The foundation date of each NGO is stated and a brief summary of its main objectives is given. These summaries – mostly of a single sentence – I have compiled from the literature presented by the organizations in Rio de Janeiro.

By contrast with India's total of a hundred participating NGOs, there were only 124 contributions from the whole of Africa. In relation to India's immediate neighbours, eleven NGOs were accredited from Pakistan (including the World Muslim Congress), ten from Sri Lanka, nine from Bangladesh and six from Nepal. The distribution of NGOs within India was fairly even, with a predictably large number based in Delhi (fifteen). The large number located in Ahmedabad (eight) is surprising. There is no preponderance of north over south (or vice versa), but only one NGO is located in the northeast (Manipur, No. 76).

Most of the NGOs were founded in the 1970s and 1980s, which is to be expected since it was during these decades that India began to take an active part in international environmental politics. We have already noted the important role played in formulating Indian national policy by the Tata Energy Research Institute, founded in 1974, and the Centre for Science and Environment, founded in 1980.

The objectives appended to each NGO are based on what was submitted to the Earth Summit. Thus the All India Women's Conference (No. 3), founded in 1927, engages in many women's activities, but from the point of view of the statement which they produced at Rio, their main concerns are tree planting, water conservation, and the setting up of training camps and seminars. Many NGOs specifically mentioned women's education, vocational training and income generation for women, especially for poor rural women and tribals. Family planning was occasionally mentioned (e.g. by All India Women's Studies and Development Organization, No. 4), but the emphasis was more often on women's overall welfare.

Considering the official attention given by the Earth Summit to climate change, and the earlier involvement of the Indian Government with this issue and ozone layer depletion, it is significant that very few Indian NGOs were concerned about either. Those which mentioned them were Balmunch (No. 9), the Gorakhpur Environmental Action Group (No. 41), the Peace Trust (No. 69) and the Tata Energy Research Institute (No. 91). One or two indicated that they would communicate whatever emerged from the summit to their constituencies, but on the whole there was a noticeable lack of interest in global warming and ozone depletion.

Most of the Indian NGOs were primarily concerned with environmental education and practical measures centred around forestry, water supplies, sanitation, energy, community health, literacy and the empowering of women and young people. Family planning, wildlife protection, pollution control, industrial safety, wasteland development, natural disaster limitation and the protection of plant species were also mentioned, but less frequently. The social and environmental dimensions of the various activities were barely distinguishable from one another.

Several NGOs mentioned Gandhian or less specific culturally appropriate ways of alleviating social and environmental problems. The Academy of Gandhian Studies (No. 1) was one such; although its main emphasis was academic, it was also involved in organic farming in rural Andhra Pradesh. The Centre for Culture and Development (No. 13) was set up to promote culturally appropriate participatory development and indigenous health practices in rural parts of Tamil Nadu.

The Gram Swarajaya Samity Bakhtuyarpur (Organization of Independent Villages of Bakhtuyarpur), based in Bihar and founded in 1978 (No. 42), ran two parallel training programmes: the one geared to grass-roots work on sanitation, air quality and the use of fertilizers without harmful side effects; the other described as 'an awareness generation programme as a spiritual and mental counterbalance'. Gramodaya (Village Awakening) in Tamil Nadu

(No. 43) believes that local knowledge must be integrated into self-help schemes. The Ladakh Ecological Development Group (No. 63) bases its ecological and sustainable development on traditional Buddhist culture, and both the Sanskrit Shodh Sansthan (No. 80) and the Sarita Society (No. 81) emphasize the use of traditional methods and resources.

The Sri Aurobindo Society (No. 88), which derives its inspiration from the teachings of the religious reformer, Sri Aurobindo, describes itself as being engaged in research in the social sciences, education, natural energy sources and Indian culture. The Tropical Botanic Garden and Research Institute (No. 92), founded in Kerala in 1979, protects endangered plant species and explores their traditional culinary and medicinal uses.

There were no signs of NGOs associated with politically rightist Hindu groups. The Brahma Kumaris World Spiritual University stated in its literature that it had submitted a case to the UN preparatory meetings, but it does not appear to have been accredited. According to its literature at the NGO Forum:

> The Brahma Kumaris World Spiritual University is of the view that the global crisis constituted by these [ecological] challenges is, at root, a spiritual crisis calling for solutions based on spiritual principles . . . We must transform the way we think, re-evaluate our priorities and principles and revive abandoned but more caring and respectful attitudes, values and lifestyles, in relation to Nature and to humanity as a whole. There will be no end to the pollution of Nature until we have put an end to the pollution in our minds . . . Scientists have already demonstrated that all parts of the world are interconnected; it is now up to the people of the world to demonstrate in their lives that we truly are one family.[4]

Two Christian organizations were accredited to the summit. These were the St Xavier's Social Service Society (No. 79), based at Ahmedabad, and the Evangelical Fellowship of India Commission on Relief (No. 36), located in Bangalore. There are many Roman Catholic schools and colleges named after St Xavier; their members have always played an active role in social service. Increasingly, some are taking part in environmental activities. It is therefore not surprising that one such group raised enough support to participate in the Earth Summit. The involvement of the Evangelical Fellowship of India – a Protestant organization which has links with the USA – is more unusual because such groups usually stress salvation out of the world rather than ways of improving it.

Taken as a whole, the hundred Indian NGOs which took part in the Earth Summit presented a closely-knit range of environment and development issues much more in keeping with the priorities of Drèze and Sen than with those of the official UN agenda. The Gandhian approach advocated by Gadgil and Guha, Sunderlal Bahuguna and others was represented, as was the 'appropriate technology' emphasis of C. P. Bhatt.

Personal initiatives

On World Environment Day, 5 June 1998, Sunderlal Bahuguna issued a message from the people of the Himalayas to the national leadership in Delhi:

> The main role of our beloved Himalayan forests is to maintain the balance in the climatic conditions of the whole of Northern India, and the fertility of the Gangetic plain . . . The findings of meteorologists and glaciologists regarding the increase in temperature and recession of glaciers cannot be ignored. Satellite imagery suggests that we now have only 31 per cent of forest area remaining in the eight hill districts of Uttar Pradesh . . . Join us in convincing our government that any further commercial exploitation of the Himalayas through the clear-felling of their forests, or the building of destructive projects such as the Tehri Dam can only result in weakening the foundations of our security for tomorrow.[5]

The statement encapsulates several arguments. Bahuguna addresses the problems posed by encroachments into the Himalayas from the experience of his earlier involvement with the Chipko movement. The forests sustain and are sustained by human communities, so that to destroy one is to destroy the other. Forests maintained in a symbiotic relationship with their natural environment are a source of good soil, pure water and clean air, and the introduction of commercially valuable species disturbs this balance. To these arguments Bahuguna now adds the likelihood of global warming and appeals to satellite technology to buttress his evidence for the depletion of forest cover. However, the final allusion to the destructive potential of the Tehri Dam does not follow logically from his earlier reasoning and begs several questions.

During the 1950s a number of large-scale river valley projects were completed without any local opposition. These included the Bhakra-Nangal Dam in the Punjab and several others. Displaced villagers were encouraged to accept their fate in return for meagre compensation on the grounds that they were contributing to nation building. As time went on, however, displaced villagers became more aware of the injustices perpetrated against them and of the truth of anthropologist Thayar Scudder's dictum that 'next to killing a man, the worst you can do is to displace him'.[6] By the 1970s organized opposition to large-scale dams was gathering force.

Sunderlal Bahuguna's ashram is situated near Tehri, a small town on the River Bhagirathi which flows downstream into the Ganges. In 1972 the Tehri project was launched jointly by the Government of India and the State Government of Uttar Pradesh to build a dam to supply electricity to northern towns as far as Delhi and to irrigate substantial areas of land. However, thousands of families would require resettlement and compensation.

The construction of the dam has been consistently opposed by the Tehri Baandh Virodhi Sangarsh Samiti (Committee for the Struggle against the

Tehri Dam), led by veteran nationalist leader Virendra Saklani. Further east former Chipko leader, Chandi Bhatt, leads opposition to a smaller dam on the Alakananda river. The main objections to these river projects are that the mountains are relatively new and subject to seismic activity and that the dams threaten the ecological balance of the region (the Alakananda Dam is close to the famous Valley of Flowers). In 1970 there was a devastating flood in the Alakananda Valley, and in 1991 a major earthquake occurred in the upper Bhagirathi region. In 1998 a large crack appeared on the mountain face forming part of the reservoir of the Tehri Dam.[7] The government continues to press on with the project.

Two more controversial river valley projects are the Silent River Dam in Kerala, which was opposed by the Kerala Shastra Sahithya Parishad, and the huge Narmada scheme in Madhya Pradesh, which will displace more than 100,000 people, mostly tribals. Opposition to this project is largely associated with Medha Patkar, though Baba Amte and Sunderlal Bahuguna have also taken part in protests. However, there are also alternative options such as diversion 'run-of-the-river' hydro-electric projects which cause very little social or environmental disruption, or the more extensive use of medium-size dams and micro-hydro projects. According to Ajay Mathur, a senior scientist at the Tata Energy Research Institute, many of the problems of the Tehri Dam might not have occurred had the construction started several years later.

Medha Patkar sometimes fasts in order to press her demands in relation to the Narmada project (e.g. in June 1993), but otherwise does not use Gandhian or Hindu methods. The scheme was finally approved by the Supreme Court in October 2000. Sunderlal Bahuguna has undertaken several lengthy fasts which, he emphasizes, are prayerful fasts at the end of which he emerges 'stronger, surer of myself. These are Gandhian ways of assertion.'[8] One such fast at the end of 1996, which lasted fifty-six days, was an attempt to persuade the government to delay blasting at the Tehri Dam until he had seen the prime minister.[9]

Several months after Bahuguna's 1996 fast, a Hindu *sammelan* sponsored by the Shankaracharya of Kanchi and the World Hindu Federation based in Nepal declared its opposition to the construction of the dam and called for measures to protect the Ganges from pollution. The three-day meeting in Haridwar was attended by senior representatives of the VHP and RSS, the former of which announced the launching of a *janjagaran* (public awareness) campaign against the construction of the Tehri Dam.[10] Another VHP dignitary declared on a subsequent occasion that 'if the government tried to stall the free flow of Mother Ganges by building the dam, I will do to it what I did to the Babri Masjid'.[11]

In contrast to such dramatic sentiments there are numerous examples of people who have inspired others to improve the quality of their natural surroundings. A New Year series of profiles in *India Today* selected fourteen 'angels of change' who have made significant improvements to their surroundings.[12] Five of these are involved in essentially environmental initiatives. I summarize their contributions:

> Nirmal, an officer in the Indian Overseas Bank, became concerned about garbage near his home in Chennai (Madras), when several poor people became seriously ill with gastro-enteritis. He started a scheme whereby the better-off residents paid ten rupees each to enable children to remove garbage with brooms, and adults (often himself) to take it away in mobile dustbins consisting of tricycle carts. 'Use citizens to sort out citizens' problems', says 'Garbage Uncle', as the children call him.

> Kinkri Devi is an elderly resident of Sangraha in the mountains of Himachal Pradesh, where limestone quarrying has reduced forest cover and firewood availability, polluted water supplies and degraded agricultural land. A widow who supports herself as a part-time sweeper, Kinkri remembers the time when the hills were covered with lush forests and sparkling streams.
> In 1987 Kinkri attended a seminar organized by a local environmental group, which inspired her to write a letter to the high court about these problems. When there was no reply she sat in front of the high court and fasted for two weeks until it agreed to take up the issues. Eventually it imposed a total ban on blasting, and in 1991 it directed the Ministry of Environment and Forests to undertake a complete study of the effects of mining on the region. Several mines have been closed and a programme of afforestation has started. 'Mineowners have money and I only have willpower. But now my battle to save the hills has been taken up by other people as well,' says Kinkri.

> Harnath Jagawat from Panchamahal in Gujarat was a personnel officer in a private company until he decided to help local people, mostly tribals dependent on agriculture and forests, to harvest water. The rains frequently failed, and even when they were adequate, much of the water was lost because the villagers could not afford to build storage dams. Jagawat and his wife set up a foundation called Sadguru which provided loans to enable people to build check dams and mini-lift irrigation schemes and to maintain them. Seeing the success of the work, the state government offered financial support, and funds also came from Norway and the European Union. During the last twenty years the foundation has made it possible to plant more than twenty million trees and irrigate 80,000 acres of land. The prize exhibit is a 110-metre long dam. 'Our plan is to ultimately wipe out poverty from the face of these districts,' declares Jagawat.

> Amulya K. N. Reddy distinguished himself as a research chemist in his early years, but he became disenchanted with the notion of hi-tech industrial progress and decided to work on a range of capital-saving, labour-intensive technologies for the rural poor. He started the Centre for

the Application of Science and Technology in Rural Areas at the Indian Institute of Science, Bangalore, in 1974. The centre has promoted inexpensive wood-burning stoves, solar ponds, brick and tile kilns and community biogas plants. 'It is the villagers who taught me to look at development from their eyes,' says Reddy. 'They say it is okay to make mistakes, unlike the people who cheer when a satellite goes up and jeer when it crashes into the sea.'

Bhandawarda Bai has attempted to restore traditional herbal medicines to the Udaipur region of Rajasthan. She did this at first by making a survey of traditional village doctors, only to discover that the availability of the rare herbs which they required had been adversely affected by deforestation. She encouraged them to upgrade their skills and teach new practitioners. On one occasion when the district medical administration failed to treat a person with malaria successfully, Bai was able to use a preparation from a plant called *nahi* to effect a cure. She has convinced many people of the value of herbs as a cheap and reliable system of medicine and of the need to conserve forests to protect rare plants.

Amulya Reddy is well known for his eco-development work. The potential of environmental legislation as illustrated by Kinkri Devi's success with the high court is an issue we shall consider later. *India Today* introduces the initiatives taken by these remarkable people by describing them as ordinary people who, having little, give all. They are described as what Gandhi would have called 'determined spirits who are fired by an unquenchable faith in their mission'. They are like '*karmayogis* who prefer to enlighten themselves through action rather than meditation. These are not human beings having a spiritual experience, [but] . . . spiritual beings having a human experience.'[13]

North India

The foregoing list can be extended further by considering other examples of people who have either been highlighted in the press or are known to me. Anuben Thakker, a primary school teacher who runs Muni Ashram near Vadodara in Gujarat, wears a saffron robe as she ministers to orphans and members of a vocational training centre. During the past few years she has introduced a biogas plant based on cow dung, a paddy de-husker and a device for extracting edible oil from groundnuts and sunflower seeds. The centre also runs a modern eighty-bed hospital, a well-stocked library and a meditation centre.

Anuben's balance between activism and meditation contrasts with the passionate Gandhian socialism of Chunibhai Vaidya, nicknamed Durvasa after the mythological fire-spitting Hindu sage. Now eighty years old, Vaidya first saw Gandhi from the roof of his house in Navsari (Gujarat) during the Dandi salt march. He wears a saffron coat, believes in voluntary poverty, and on one occasion turned down a gift of a jeep from an admirer. It is for his temper, however, that many people know him best. In 1984 Vaidya confronted the Gujarat chief minister to remonstrate with him over the tardy progress of drought relief. 'If things don't come around, I will thrash your district collector with *chappals* (sandals),' he threatened! He believes that the Narmada Dam is an essential source of water for farmers in his area. 'We need the dam. There's no alternative to it,' he declares, defiantly.[14]

Another Gujarati, Shyami Antala, has revolutionized water conservation following the four-year drought in Saurashtra between 1984 and 1988. He began his career as circulation manager for a Gujarati daily newspaper. His first attempts at reviving wells met with indifference and suspicion, so he went to discuss his ideas with the leaders of two Hindu sects, the Swaminarayans and the Vaishnavas. They took up Antala's ideas with enthusiasm, and groups of saffron-clad ascetics spread the message that if nothing was done to harvest water supplies Saurashtra would be a desert in thirty years. The people listened, and Antala's advice was heeded, with the result that a quarter of a million wells have been made operative. Antala now wants to promote his ideas all over India: 'I feel like going all over and telling [people] that they should forget about the government and what it is not doing for them. Be self-reliant. Do it yourself. You can solve your own problems.'[15]

Antala was fortunate in being able to enlist the support of the Maninagar branch of the Swaminarayan sect (one of the four which follow the teachings of the nineteenth-century reformer, Swami Sahajhanand). In October 1997 this branch held a world conference in Allahabad which gave special consideration to environmental problems, earmarking sums of money for desert areas in the Kutch region, where many of the sect's well-off overseas members originate. According to Purshottam Priyadasji, the sect's current head, 'We believe in converting the devotees' religious fervour into positive action for society.'[16]

The Swadhyay movement is also strong in the Saurashtra region of Gujarat. This is a religious organization based on the philosophy of Panduranga Shastri Athwale, usually known as Dadaji, who believes that strong faith in God is the basis of all uplift. *Swadhyay* means study of the self, and is common to all religions. According to the *swadhyay* approach, poverty is related to our state of mind and leads to self-limiting social, economic and religious practices. Labour is the most powerful form of devotion to God, and everyone should try to offer two days of labour every month as a form of worship. Villages in which a sizeable proportion of people support *swadhyay* have started to act as centres for agricultural research and sustainable technology. Chemical fertilizers and pesticides are discouraged, and there is increasing reliance on biomanures and indigenous pesticides. It is difficult to estimate the current size of the movement, which claims to operate in twelve states.[17]

India's most populous state, Uttar Pradesh, contains some of the most important Hindu centres, including Vrindaban where, according to legend, the Lord Krishna grew up. During the 1980s a retired engineer called Sevak Sharan became so concerned by the felling of trees in Vrindaban that he founded the Vrindaban Conservation Project to try to stop it. In 1991 this became the Vrindaban Forest Revival Project in association with the World Wide Fund for Nature, and two years later links were established with the Hindu community in Leicester in the UK. A regional coalition called the Vraj Seva Mandal is working on a programme of environmental planning, sanitation, the restoration of culture and the protection of sacred groves, which are important centres of biodiversity in this area and where abundant wildlife and water resources are to be found.[18]

In the more mountainous districts of Uttar Pradesh, in the Garhwal Himalayas, Arun Kumar Badoni, the grandson of a renowned *vaid* (traditional healer), has grown and encourages local villagers to grow a variety of rare and endangered plants. These medicinal and aromatic plants include *ateesh* (*Aconitum heterophyllum*), which is in great demand both in the Indian Ayurvedic industry and by foreign herbal pharmaceutical manufacturers. After trying unsuccessfully to alert government officials to the over-harvesting of rare Himalayan plants by big business, Badoni encouraged villagers to grow their own and market them locally. He founded the Society for Himalayan Environmental Research in 1989 to promote his ideas. More recently a field research station has been set up in the Tons Valley, and a temple in Kedernath has started to cultivate herbal plants.[19]

Before moving south, I will mention the work of Sukhbir Tanwar, a former sub-inspector of the Central Reserve Police Force in Delhi. Tanwar established the Utthan Vahinee group to collect garbage in Vasant Kunj. For thirty rupees per household per month, Tanwar and ten assistants collect garbage, transport it in a three-wheeler cart, and dispose of it on unused land owned by the Delhi Development Authority (with its permission). Recyclable materials are separated and sold. Biodegradable waste is kept in a pit with phosphate fertilizer, salt and four inches of soil. After three months the resulting manure is given to the residents to grow vegetables. The venture had provided paid work for several previously unemployed people.[20]

Kiran Bedi, who was Inspector-General for Prisons, with responsibility for Tihar Jail in Delhi, also introduced income-generating forms of garbage disposal. Her reforms were so radical that the entire prison complex, which included special buildings for meditation, was renamed Tihar Ashram. Her radical ideas and personal charisma upset the Delhi political bosses, however, and she was transferred.

South India

Andhra Pradesh is a state where there have been several innovative development projects. The Janmabhoodi Programme, launched by the state government in 1997, includes all the elements of participative economic and environmental development noted earlier. Village assemblies designate essential projects, and people execute them via a combination of *shramdaan* (voluntary labour), professional expertise and government funding, usually in the form of loans. According to the chief minister, Chandrababu Naidu:

Environmental conservation is our main concern and our endeavour . . . [We will] ensure that there is no water stagnation in the villages and slums; that all drinking water sources are repaired.[21]

Communities are involved in collecting garbage and treating organic waste through vermiculture. Rainwater harvesting is promoted, and women's self-help groups based on specific interests such as credit, forest protection, and waste land and watershed development are being organized under the guidance of social motivators. Critics accuse the state government of political opportunism and the duplication of existing channels for social welfare, but the public response has so far been positive.

On a much smaller scale, Darepalli Ramaiah, a Khammam potter, cycles round his locality distributing saplings. 'As you write or chant Ram ten million times for *mokṣa*, plant ten million saplings for a pollution-free world,' he urges his listeners. Ramaiah has personally planted 70,000 saplings and spoken about tree protection at 300 schools, 170 offices and 162 temples in Andhra Pradesh. Everybody knows of his love for trees, and government officials regard him as a useful ambassador for their environmental outreach.

Ramaiah's passion for trees goes back to his schooldays, when he first learned how trees give oxygen, shade and fruit in return for water. He populated his family's farm with a wide variety of trees, some familiar, others less well known, distributing them at weddings and on other occasions with a few words about their medicinal, nutritional or other values. Children know him by various names: *Vruksha Pitha* (father of the trees) and *Vana Jeevi* (protector of the trees), for example. 'Nature takes care of us only if we take care of it,' he says.[22]

Mention might also be made of K. Azariah, another grass-roots development worker in Khammam, associated with the Church of South India, who has done a great deal for the poorest sections of villages in his region.

Moving further south to Mysore's Bilgiri Rangana Hills, there is an inaccessible community of Soliga tribals who worship their local doctor, Hanumappa Reddy Sudarshan, almost as a god. Sudarshan decided as a young medical graduate that he would work in the Nilgiri Hills among Toda and Phaniya tribes with another doctor. In 1980 he moved to the Soliga area, where he was at first regarded with fear until he cured a victim of snakebite. He was soon inundated by villagers who had been trampled by elephants or mauled by bears. In time he was able to introduce preventative medicine and set up a school where the children received a modern education but were also taught about their own tribal customs: 'In my school I have children who can identify 240 species of tree by name,' he boasts.

In addition to building up the local community, Sudarshan has been obliged to confront local landlords who tried to take over tribal land. On one occasion he challenged the notorious sandalwood smuggler, Veerappan, after he killed four Soliga tribespeople because he thought they were police informers. 'If you are so bloodthirsty, you should take my life first,' Sudarshan declared.[23]

The environmental problems of the Nilgiri forests are well illustrated in *Cheetal Walk* by E. R. C. Davidar, a vivid picture of life in the Sigur Reserve on the Tamil Nadu side of the mountains.[24] This is a well-informed account of wildlife protection in the southern mountain ranges. Naturalist Valmik Thapar has also raised public awareness about the need to protect tigers. He believes they form the apex of the food chain and that saving them helps to preserve our total habitat.[25]

In Appendix B the participation of the Bombay Natural History Society, established in 1883, at the Earth Summit in Rio de Janeiro is noted. The society was closely associated during the 1960s and 1970s with Salim Ali (1896–1987), who was probably the most outstanding biologist and conservationist of twentieth-century India. His major work is on birds of India and Pakistan.[26]

In the southeastern state of Tamil Nadu there is a shy Tamil nun who has made a discovery that could revolutionize the technology of shipping. Sister Avelin Mary, principal of St Mary's College, Tuticorin, is a marine biologist who identified a coral extract called *Juncelia juncea* which inhibits biocorrosion on the hulls of ships. The discovery has yet to find commercial backing, but in 1998 the Biographical Institute in the USA named Sister Avelin Woman of the Year. 'Our research is equal to the West. There is no magic in the world except the magic of hard work,' she said.[27]

Moving to the extreme southwest, to Kerala, there are a number of participative development programmes which address environmental issues. Whereas the central government under Rajiv Gandhi had called for *panchayat*-level planning, the Andhra Pradesh Janmabhoomi Programme side-stepped the *panchayats* by appointing *gram sabha*s (village assemblies) convened for one week. In Kerala the Kerala Shastra Sahithya Parishad had encouraged grass-roots planning since the 1970s, but it was not until 1990 that the State Planning Board allocated funds to local bodies.[28]

During the last few years *panchayats* in Kerala have promoted wells and water tanks, medicinal plants have been studied and cultivated, and biofertilizers have made it possible for some farmers to grow paddy twice a year for the first time. Villagers in the Kannur district have even managed to build a cost-effective micro-hydro project that will generate four megawatts of electricity.[29] However the programme as a whole is hampered by bureaucracy and political rivalries.

The problems of fishing communities in Kerala do not fit into the mainstream understanding of development, and the communities are widely scattered. During the 1980s large ships were exploiting waters reserved for traditional fishermen, mechanized vessels were using bottom trawling that damaged the sea floor, and licences were given too readily to large foreign ships. A campaign by one individual put a stop to all this.

Thomas Kocherry trained as a lawyer and became a priest, joining the Redemptorist Congregation because he was inspired by its founder, St Alphonsus Lagoury, who left a lucrative legal job to work among poor shepherds in Italy. As a priest in a poor fishing village near Thiruvananthapuram, Kocherry soon became well known for his support for the rights of his people. When the police arrested him, he fasted, while the fisherfolk waited day and night outside the police station for his release. When they expressed their anger for many injustices by turning away from their traditional political party, the Congress, to support the communists, the Church blamed Kocherry and transferred him to Bangalore.

As a result of Kocherry's protests, the Marine Regulation Act, designed to keep large vessels outside a 22 kilometre limit, was eventually enacted; there were also other gains for the fishing communities. Kocherry has become an international figure, but fame has not compromised his ideals, and he turned down an award from the USA for his work on the grounds that the oil company which funded the award was responsible for marine pollution:

> I am grateful to . . . Greenpeace who nominated me for this award . . . [but] the heirs of yesterday's polluters are becoming the allies of today's environmentalism. This is hypocrisy.[30]

It is an interesting gloss on this episode that in presenting an essentially complimentary profile of Kocherry, *India Today* records that a financial award from the USA was accepted by him, without considering the possibility that he turned it down![31] In June 1999 he was offered the prestigious Norwegian Sophie Prize, which he accepted.

Points of view

I have described the work of a number of people engaged in environmental activities and now consider various points of view. The material contained in this section is an abridged version of interviews conducted between 1998 and mid-2000. Each interviewee was asked a series of questions about the ways in which the Hindu tradition can increase environmental awareness, the manner in which the unecological story about Krishna and Arjuna in the Khāṇḍava forest should be understood, south Asia's major environmental problems, and reactions to the views of Amartya Sen and the work of Sunderlal Bahuguna and others. The responses are presented as a continuous narrative. Amartya Sen asked to present his ideas according to a different format.

Karan Singh

We begin with Dr Karan Singh, a distinguished scholar and statesman, whose summary of the basic tenets of the Hindu tradition was included in Chapter 2.

Karan Singh was heir apparent to the princely state of Jammu and Kashmir, but renounced his privy purse and embarked upon a political career which brought him into Mrs Gandhi's first Union Cabinet in 1967 at the age of thirty-six. He is currently President of the People's Commission on Environment and Development and has campaigned vigorously to make successive governments aware of the need for a healthy environment. What follows is a conflated summary of two interviews:

> The Hindu tradition can play a large part in increasing environmental awareness through education and to a lesser extent the media. In education this should start at the secondary level.
>
> In our cultural heritage the environmental values are extremely strong. The Atharvaveda, for example, contains the sixty-three-verse Bhūmi Sūkta (Hymn to the Earth). These verses provide the most integrated, the most enlightened and the most realistic set of thoughts on the environment to be found in any literature anywhere in the world. They express a world view which recognizes the spirituality inherent within nature and stresses that, to survive, the human race must partake of the holistic and harmonious relationship with the world that it inhabits.
>
> Through the medium of education we must stress the parts of the Hindu tradition that give a sense of sacredness to all nature: mountains, rivers, the Himalayas. Young people should learn that there is divinity in all nature. They should engage in massive programmes of tree planting. K. M. Munshi had the splendid idea of *vana mahotsava* (festival of trees). We should restart this. Each temple should give saplings to plant instead of the usual *prasad*. Temples used to have more land, and flowers to be offered to the deities were grown around them. Temple gardens were used to grow Ayurvedic plants; they were the deity's kitchen garden. We should encourage religious leaders of all traditions to say that there is religious merit in cleaning the environment – for a Hindu or Buddhist that this is good karma.
>
> Our sages and seers stressed the sacredness of rivers, lakes, land and forests. They realized in their higher consciousness that unless we are able to protect the natural environment, the human environment will collapse. We should restate these ideals in a new form in the context of our present-day technologies and make them more meaningful within our contemporary framework. We need to incorporate environmental values in our formal and non-formal educational systems. We need a coordinated and orchestrated programme of public awareness through every available medium to build up public opinion to renew our ancient concept of living in harmony with the resources of our planet.
>
> Episodes such as the destruction of wildlife by Krishna and Arjuna should be ignored. It reflects the change from the early stages of history to settled agricultural life. Elsewhere Krishna is a very pastoral person who is to be found among the cows. The uncharacteristic episode may be

an interpolation. But in any case scripture is of two kinds, what is declared in the Vedas and what is remembered and is evolving. This episode with Krishna and Arjuna is not part of the most authoritative scripture. We are entitled to extrapolate from the teaching of modern seers as well. For example, Sri Ramakrishna often referred to the existence of Śiva in all selves, which means that every time we serve others we serve God. But I am sure that today he would say that when we serve the needs of our physical world we are also serving God. Creative extrapolation is an important reason for Hinduism's resilience; I prefer *sanātana dharma* to the word 'Hinduism'. The Hindu scriptures teach that God is within people and nature – that there is a community of well-being.

India's main environmental problems are to do with deforestation, the lack of political will and the pressure of overpopulation. There is gross overconsumption in the West which must be reduced, but that does not excuse India from taking steps to reduce levels of population increase. As chair of the Indian Board of Wildlife I encourage whatever steps can be made to improve the quality of our environment. However, since the departure of Mrs Gandhi from office little has been achieved politically.

Amartya Sen is correct to stress the basic needs of life: education, healthcare and the empowerment of women, but he does not say enough about the population issue. The cancellation of international debts is not a major issue as far as I am concerned. I am not an economist. We need trade rather than aid, not charity. We need assistance with low-pollution technology in various industries and transport. We must tackle pollution right across the board. I am very impressed by the way the River Thames has come back to life in Britain. We must do the same for the Ganges. Could Britain help us with that? It would provide great spiritual merit! If countries such as Britain want to assist India then they should concentrate on one specific area.[32]

Karan Singh's views as recorded here are consistent with his published opinions as summarized in Chapter 2. He is more emphatic than many people of the need for south Asians to reduce the rate of population increase, though he tempers what he has to say by putting it within the context of the empowerment of women and overconsumption in western countries.

M. C. Mehta

M. C. Mehta is a Supreme Court Advocate and internationally acclaimed environmental campaigner, who has been largely responsible for stimulating the 'judicial activism' of the Indian courts' major environmental rulings. We must digress briefly to consider certain aspects of the legal system.

The Indian constitutional pattern resembles the British one in many respects, but differs in that the Constitution itself is supreme, and Parliament is subject to it. This means that the Indian courts can adjudicate on the constitutionality of any law passed by Parliament. When the courts exercise their power of judicial review in an interventionist manner it is customary to speak of 'judicial activism'.

During the 1950s and 1960s the courts challenged Parliament over several incidents concerning the fundamental rights of property and trade, and in 1970 they invalidated Mrs Gandhi's Bank Nationalization Act on the grounds that no provision was made for compensation as required by the Constitution. During the 1975 Emergency, the courts did little to protect the fundamental rights of individuals, but from 1977 onwards there was 'a sea change in the judicial mood and temper'.[33]

Two landmark judgements by the Supreme Court led to a redefinition of 'life' in Article 21 of the Constitution to include basic human necessities such as pollution-free air and water, and the liberalization of the notion of *locus standi* such that breaches of fundamental human rights could be taken up by others. The latter led to Public Interest Litigation which, combined with the former and two environmental Amendments to the Constitution in 1976 and 1977 (Articles 51A and 48A), paved the way for some major environmental directives by the Supreme Court. M. C. Mehta has played a major part in the Supreme Court's environmental litigation:

> Religion generally and the Hindu tradition in particular has an important part to play in raising awareness of environmental problems. The pollution of the Ganges at Haridwar is so serious that on one occasion the river caught fire. Just imagine it for a Hindu – the holy river itself was ablaze! This type of imagery is powerful in helping people to care for the natural world as sacred. All religions can evoke such feelings. The worship of rivers is part of our civilization and culture.
>
> When Hindus worship we do so not just with an attitude of devotion. We worship by our actions, as the Bhagavadgītā teaches. We must therefore keep the water in our rivers clean and pure. It is our custom to put some water from the Ganga [Ganges] into the mouth of a young child; shortly before death we also like to receive Ganga water. Ganga *jal* – it's ingrained in the minds of Hindus. Our ashes are immersed in the Ganga and the rituals are carried out at Haridwar. So it's shocking to us to find how polluted the Ganga has become. We must respect our rivers and not pollute or put garbage into them. The West appreciates the need for clean water, but it does not have any deep feeling for rivers as we do. Also fish may be eaten, but should be given respect and not tortured senselessly . . .
>
> I do not get my Hindu beliefs from reading the scriptures, and there are many passages that I am not familiar with. I do not know the story of Krishna and Arjuna razing the Khāṇḍava forest and flinging all the escaping animals back into the flames, for example. My beliefs come from my elders (who may have read more of the scriptures than I have), and from my experience.

Trees are sacred to us – *tulsi*, *pīpal*. When I was a child I used to see women moving in a circle round a tree each morning, but I didn't understand the reason. Now I do. I understand the scientific reasons underlying those beliefs. The *pīpal* tree gives out oxygen continuously for twenty-four hours. Such scientific knowledge was put into a spiritual form by our ancestors. The forest cover was preserved and especially those trees which we now know give out oxygen continuously. Our attitude to trees was governed by custom and moral sanctions, but people came to disregard these.

The opinions of Gadgil and Guha about our environment are helpful, but in some respects they are quite western. We must not ape the West... Amartya Sen's analysis is preferable, but we need to give more attention than he does to population growth. An extra third of the population can absorb all our improvements in standards of living.

Sen's view that we need more primary and secondary education is correct. But what kind of education? Instead of 200 universities we should have more need-based and vocational education. The whole pattern of education should be changed.

It remains to be seen what the BJP will do at the political level to improve things. They may have a good environmental agenda, but there's a big difference between preaching and practising. All the political parties preach more than they practise, which is why our natural resources are so depleted.

Gandhi was shrewd in his espousal of village republics, *panchayat raj*, and grass-roots activities generally. Ultimately this gives more power to people and especially to women. Religion need not be detrimental to women's participation. In all our family rites my wife must take part otherwise they are not holy. This belief goes back to antiquity. In religious ceremonies husband and wife are the right and left hand...

We cannot proceed according to a single global model of materialism and consumerism. India must not imitate this model and we must all change. People can be very happy in their traditional surroundings. We must shun materialism. We must minimize greed and maximize our response to need. Our needs are important, our greed is not. Every child in the USA consumes the food and proteins of twenty children in India.

We must avoid the plastic culture of the West. The USA is a throwaway society. We must encourage indigenous systems and not purchase things manufactured outside. Chemical contaminants such as pesticides are not necessary; they contaminate water and blood and get into the food chain. We must learn to conserve our own resources. Gandhi patched and repatched his clothes. The Buddha taught his monks not to throw cloth away. Hindus traditionally practise conservation... Religions have a lot to contribute to the solution of environmental and social problems, but ultimately there is only one true 'ism', and that is humanism.[34]

Amartya Sen

Professor Amartya K. Sen is a Nobel Prize-winning economist and the Master of Trinity College, Cambridge. I have summarized his development-led views on social and environmental problems in Chapter 7. His approach to religion is more philosophically thoroughgoing than either Karan Singh or M. C. Mehta. In an essay written in April 2000 entitled 'Consequential Evaluation and Practical Reason', he challenges both the conventional interpretations of the Bhagavadgītā's dialogue between Arjuna and Krishna and Gandhi's refined version of it. As the rival armies face each other, Arjuna sees his own kin on both sides and decides that he cannot kill them. Sen comments:

> Krishna points to Arjuna's duty to fight, irrespective of his evaluation of the consequences. It is a just cause, and as a warrior and a general on whom his side must rely, he cannot waver from his obligations. Krishna's high deontology has been deeply influential in moral debates in the subsequent millennia. It is, I suppose, a tribute to the power of pure theory that even Mahatma Gandhi – no less – felt deeply inspired by Krishna's words on doing one's duty irrespective of consequences, even though the duty in this case was for Arjuna to fight a violent war (not in general a cause to which Gandhi could be expected to warm) . . . In this essay, I intend to take the other side – that of Arjuna – and proceed from the basic idea that one must take responsibility for the consequences of one's actions and choices, and that this responsibility cannot be obliterated by any pointer to a consequence-independent duty or obligation.[35]

Sen proceeds to discuss two basic levels of consequential evaluation, the broad characteristics of a specific consequentialist approach, and the need for situated evaluations such as Arjuna's recognition that he himself will have to kill some of the people he feels affection for: 'This contrasts with the utilitarian formula that the evaluation must be, in every way, independent of the evaluator, and in particular, must take the very specific form of maximizing the sum total of utilities.'[36] It does not, however, 'violate any requirement of "impersonality" that may be imposed on the discipline of ethics'.[37]

Sen discusses some of the traditional problems of consequentialist ethics, such as the comparison of alternatives and the concept of comprehensive outcomes. Human rights may need to be balanced against one another or against other good things. How, for example, do you decide whether or not to steal a car in order to prevent a murder? He comments:

> The need to examine the relative importance of different consequences (including well-beings, freedoms, rights, etc.) which may compete with each other in evaluative assessment does arise in many different contexts. We live in an interdependent world in which the realization of our

respective freedoms interconnects in a variety of ways, and we cannot treat them each as an isolated island. The discipline of consequential evaluation forces us to take responsibility for our choices since our actions influence other people's freedoms and lives as well as our own. The reach of our responsibility includes asking certain questions, such as those concerning the relative importance of different rights or freedoms the realization of which may impinge on each other.[38]

The placing of the interpretation of the Bhagavadgītā within the wider context of consequential versus deontological ethics lifts our earlier ecological discussion of the burning of the Khāṇḍava forest onto a different plane. For if we cannot accept the orthodox ethical interpretations of the Bhagavadgītā as summarized, for example, by Roderick Hindery, then we are left with a religious text which may inspire great devotion, but is of little ecological value in any other respect.[39] As we indicated earlier, however, all religions are multi-faceted and do not stand or fall on the basis of their religious texts.

On more specifically ecological issues, Amartya Sen was concerned that he had been accused by the editor of an environmental magazine based in Delhi of failing to understand the notion of 'ecological poverty'.[40] He denied the accusation:

> Poverty is deprivation of human freedom; among the factors that can make us substantively less free there are a number of different considerations including lowness of income, lack of basic education, lack of arrangement for healthcare, on one side; but also on the other an adverse environment, a depleted natural resource base, foul air to breathe, or polluted rivers to get your water from. So ecology comes into the notion of human poverty in a big way. It's not a notion of *ecological* poverty as such; it is part of the *general* notion of poverty. People might sometimes complain that I try to integrate the idea of ecology with the idea of development. If that is the charge, I plead guilty. I do intend to do it, and for exactly the same reasons for which Gro Brundtland in the Brundtland Report made the basic concept of the environment integral to the notion of sustainable development. Sustainability on the one side is referring to its environmental viability (among other things), and development on the other hand is the motivating end product which is the motivating factor in making us take an interest in the environment. I see no conflict whatsoever in taking such a broad view of development and ecology, and at the same time believing, as I do, that I take ecology extremely seriously.

On the issue of educating people about ecology and the need for people to acknowledge their responsibility for causing pollution and depleting resources, he had this to say:

We need to make the nature of public discussion more sensitive to ecological concerns. That's not *just* a matter of formal education in school. There are many important subjects which are not taught in school. How to be good citizens, how to decide which political party you want to vote for, are not subjects which are part of the school curriculum or for that matter the college curriculum, but they are very important. Some of the things that are very important for our lives come from formal education; others come from our participation in public discussion, from reading newspapers (for which literacy is important), from listening to news on the radio or watching it on television (if one is lucky enough to have a television or to know someone who does have or to have a communal television to which one has access). I think that to take part in public discussions in meetings is a very important educational feature of a democracy. The problems of ecology are somewhat of that kind: the fact that we can *do* something about the environment, the fact that we don't have to breathe foul air just because there are uncontrolled emissions now. These recognitions are very important because the problems are ours but the remedies are also in our hands. That is a very important lesson . . . Undoubtedly there is a role for the school curriculum in making people aware, but it has to go well beyond that . . .

We must all acknowledge our responsibility across the globe for fouling the air, for example. One way of thinking about it is this, that in the context of the population problem there is a great deal of concern in the West that the population growth rate is very high in many Third World countries today. And yet you know that even though the increase in the Bangladeshi population may be very much larger than that of the British population today, we see from the last report published . . . by the Royal Society that the additional fouling of the air generated by the additional British population is larger than the additional fouling of the air by the larger increase in the Bangladeshi population.[41]

Sacred groves

Reverting to my earlier concern about the role of the Hindu tradition in promoting environmental awareness, I include the views of a conservationist who finds great devotional inspiration in the Hindu scriptures (especially the Bhagavadgītā), and an environmental activist who was able to demonstrate the extent to which the wisdom of the past can illuminate contemporary problems.

Mrs Neelam Dewan is an expert on the fauna and flora of the Himalayas. She lives at Bishop Cotton School, near Shimla in Himachal Pradesh. Her views about the Hindu tradition and the environment are as follows:

> The Hindu scriptures are vast and contain a great deal about such things as holistic medicine, the construction of buildings and other practical

matters which are environmentally related. The scriptures are not entirely consistent. In one place the Mahābhārata may show Arjuna and Krishna throwing wildlife back into the flames of the forest, which is ecologically irresponsible, whereas in another passage Arjuna refuses his next cycle of birth unless his dog can go with him! In yet another part of scripture – the Laws of Manu – all kinds of creatures are rescued from a great flood. According to the Bhagavadgītā there is no such thing as an inanimate object and even a stone has animate properties. We must not be cruel to animals and according to the Jains we must not harm any living thing.

Hindus pay special respect to trees and rivers. We worship the *tulsi* plant, for example, and as the lady of the household I should water one every day. But unfortunately *tulsi* doesn't grow in the Himalayas! The *panchavati* is a group of five holy trees which if found together – even some of them – constitute a sacred grove. All the fruits or leaves of these trees have special uses or properties. They are *bel* (the fruit is medicinal), *banyan,* a variety of fig, *pīpal* and *neem*. We revere rivers, but here there is a paradox because we both worship and pollute them. We collect water from the Ganga and keep it in bottles. Our holy lake at Mansovar is now in Tibet. The sources of the Ganga and Yamuna are holy.

The Gītā encourages worship of the celestials. There are gods corresponding to all the elements, but there is only one universal God and all others are manifestations. There is a life force at every level: trees, birds, etc. I like to meditate on these things and to do this I need to spend time on my own. As the wife of an army officer, I have often been by myself. I grow plants in flowerpots and I spend quite a lot of time looking after them. Our son once had an argument with his friends about the wealth of each one's parents. One boy said to him: 'Your parents must be the richest because they have the most flowerpots.' Jagadish Chandra Bose has written a beautiful book about his research on electrical responses in plants.

I read a small portion of the Gītā every day. Today I have read the part in Chapter 10 where Krishna says: 'I am the glory of the Maruts, the elements, among the mountains I am Meru. Among the immovables I am the Himalayas. Among the rivers I am the Ganga.' It will keep coming back into my mind during the day.

There is little environmental concern in politics. Maneka Gandhi went overboard. Sunderlal Bahuguna worked well with tribals in forest areas but now he seems to be against progress.[42]

I visited a sacred grove in the Pune region of Maharashtra with Vijay Paranjpye, a distinguished economist and environmental activist. The sacred grove, usually known as *dev-van*, or forest of God, is located on a model farm on a tributary of the Ramnadi river between the villages of Bhugaon and Baudhan.[43]

According to Paranjpye, sacred groves are typically areas open to the sky where the land, water and vegetation coexist. Whereas villages are located in clearings, groves are to be found in the forests usually about a mile away. They are repositories of plants and animals where important social functions are performed, e.g. village headmen are installed and disputes are settled. The groves are pre-Aryan in origin and were initiated during the hunter and semi-nomadic period, well before the advent of settled cultivators.

Most village groves would have been presided over by a priest, most probably a mixture of social functionary and medical practitioner (not a brahmin because this was before the Aryan period). The grove itself was a sacred place and certain kinds of behaviour such as collecting firewood were totally prohibited. No doubt the priest told stories about the terrible things that happened to people who offended the gods by breaking the rules.

This grove was situated around a *kund* or sacred pond, which had once been used by cattle and feral animals. In the trees on the bank of the pond was a group of flat stones daubed with red vertical lines. These represented the *satmauli*, seven fierce female deities who protected the grove. A cobra lived somewhere near the stones. Twice a year the local villagers would sacrifice a chicken or a goat to appease the *satmauli*.

A short distance from the grove, there are several terraces on a hill where a search was being made for underground water. A hydrologist from Pune University took subsoil measurements of electrical resistance to test for underground streams. After an exhausting two hours of measuring soil resistivity in all directions and another hour of calculations and mapping, it was concluded that water for a wide but fairly shallow well could be found in a slightly forested area at one point on the terraces, and more water could be located at a much deeper level at another spot.

As we prepared to move back to our vehicle, Vijay Paranjpye drew my attention to the pointing branches of a tree directly above the first site, and a termite's nest on a small rock directly above the second – precisely where the Arthaśāstra, written 1300 years ago, would have located what the best modern science had taken three hours to find!

We have considered social and environmental initiatives ranging from the contributions of Indian NGOs at the Earth Summit to the efforts of small groups and individuals located in various parts of India. Where feasible, I have indicated the explicitly religious characteristics of these groups and organizations. We have noted the differences between the environment and development agenda of the United Nations, with its emphasis on macro-issues such as enhanced global warming, and the more regional and local concerns summarized in Appendix B.

It will be clear that environmental issues in India are experienced, and therefore must be resolved, within particular social contexts and along a

158 *Religion and Ecology in India and Southeast Asia*

much broader front than is usually perceived in the West. It will also be clear that the social and economic contexts of whatever solutions are feasible are in line with the proposals set out in the last chapter, especially by Drèze and Sen (i.e. the need to improve primary and secondary education, healthcare and the status of women).

I rounded off the data with a series of interviews designed to demonstrate some of the realistic possibilities for social and environmental improvement, and ways in which the Hindu tradition can contribute to them.

9 Expanding our horizons

Not only has the scope of this study been extensive in a geographical sense, but it has also been broad, since it has tried to cover both the historical and the contemporary manifestations of religion. I have also emphasized the importance of acknowledging the full range of the various levels at which religion operates – social, ethical, ritual, doctrinal, etc. – as indicated in Chapter 1.

Of course, I have not been able to give a comprehensive account of all this, and the material has necessarily been selective. The geographical coverage, for example, has virtually omitted certain areas such as Sri Lanka. Historical material has been essential in order to illustrate the significance of the continuities and transformations between the past and the present, but the coverage as a whole has been biased in favour of the present. While not overemphasizing the doctrinal and ethical dimensions of religion, I have nonetheless paid a good deal of attention to the religious texts.

I have attempted to present a backcloth against which the potential of two major religious traditions in two regions of south Asia for addressing social and environmental problems can be set – a large, but inevitably incomplete tapestry. It remains for others to fill in the gaps.

Priorities and perceptions

Before considering the potential contribution of religion to the amelioration of social and environmental problems, I challenged a number of popular views about what these are. They are not – and I gave reasons in Chapter 1– global warming and a Third World population explosion, although aspects of these may need to be taken seriously. They are much more to do with the concerns of the hundred Indian NGOs which took part in the Earth Summit and are listed in Appendix B.

We noted in Chapter 7 that a report on greenhouse gas emissions produced by developing countries published by a well-known US-based institute was demonstrated by the Centre for Science and the Environment in Delhi to have been a deliberate overestimate. Even more recently (February 2000), an environment correspondent of a reputable British newspaper made the dramatic claim that as global warming increases and the Himalayan glaciers begin to melt, north India's rivers will dry up. The political message was clear: 'Reduce your greenhouse gases, India, or your agricultural irrigation system will collapse.' The scientific argument is completely flawed however. The Himalayan rivers are fed by monsoon rains and melting snow, both of which might conceivably *increase* as a result of global warming.[1]

There is, of course, a serious population crisis. In essence – as noted in Chapter 1 – it is that every child born in the West consumes on average eight times as much of the earth's resources (e.g. in terms of energy) as a child born in a developing country. Western levels of consumption are mainly responsible for both the historic and continuing depletion of natural resources with consequent pollution all over the world, but especially in developing countries, and this is the primary reason for the increase in greenhouse gases that has become apparent during the past few decades.

From the perspective of south Asia I characterized the most serious environmental problems as loss of forest cover and biodiversity, the contamination of rivers and water scarcity, and natural resource depletion and pollution. Dr Cécile de Sweemer, a Belgian healthcare expert working in Laos, ranks the relative seriousness of Laotian environmental problems in her region:

1   Shortage of water for drinking and irrigation.
2   The need to diversify food crops (more legumes and fruits required).
3   The need to reforest with a variety of trees suitable for firewood, fodder and agroforestry.
4   Appropriate environmental education, especially for the very young.[2]

Clearly, environmental problems cannot be separated from their social context. With regard to global warming, de Sweemer's only concern is that it will increase the evaporation of water, so more will be needed, and it will inhibit the growth of staple food crops if it rises by more than about 3°C. This will take the best part of a hundred years, however, by which time temperature-resistant crops can be developed through genetic modification. In Laos there are more than 300 varieties of rice and considerable potential exists for the development of hardier new protein-rich strains which are resistant to temperature fluctuations.

I have chosen Laos as an example, but the general picture is much the same throughout the regions that we have been considering, and, as argued in Chapter 7, solutions to social and environmental problems will only succeed within the overall framework of improved primary and secondary education, healthcare, and a better deal for women.

Religious misconceptions

In the introductory chapter I indicated that the majority of young Hindus and Buddhists consider themselves to be religious to an extent that goes beyond membership of a particular community. Thai Buddhists focus their aspirations on the *saṅgha*, with its rites and festivals, and the possibility of entering into its most intimate workings through ordination, if not as a monk, then at least as a lay nun (*mae chii*). Hindus can express their commitment in a variety of ways: prayer, meditation, pilgrimage, austerity, membership of a religious order, and so on.

There is, however, an aspect of religious experience that is more difficult to chart, but which is characteristic of many young Asians who may sit loose to the more conventional expressions of religious commitment. This has been brought out in a number of novels, of which, from the point of view of the Hindu tradition, the best is Raja Rao's *The Serpent and the Rope*. Published in the 1960s, its style may now seem dated, but it nonetheless remains as deserving as ever of E. M. Forster's tribute as 'the best novel in English to come from India'.

The Serpent and the Rope – the title echoes Śaṁkara's twin allusions to reality – tells the story of Rama, a young Indian, and Madeleine, a French girl, who meet at a French university in the late 1940s and fall in love. In the course of their romance, they visit London. Rama is the narrator:

> History and my mind vanished somewhere, and I put my arms round that little creature – she hardly came to my shoulder – and led her along alleyways and parkways, past bus stop, bridge and mews – to a taxi.
> 'Let's go to Soho,' I said, and as I held her in my arms, how true it seemed we were to each other, a lit space between us, a presence – God. '*Dieu est logé dans l'intervalle entre les hommes*', I recited Henri Frank to her.
> 'Yes, it is God,' she whispered, and we fell into the silence of busy streets.[3]

Such religious sentiments may seem incompatible with the brash secularism of public life in India or the strident fundamentalism of the 'saffron brigade', but they are integral to the lives of many people, and as such are a potent and enduring part of religion, even though we cannot easily quantify them.

What we can quantify I have set out within a matrix which recognizes the multi-faceted dimensions of religion. Unlike fundamentalists in most religious traditions, I attribute no more significance to particular doctrinal texts than to so-called 'folk' religion, which often expresses the religious feelings of ordinary people at a very deep level. This may, of course, also be distorted, as when Hindu nationalists evoke the symbolism of Ram without any concern for historical accuracy. However, most fundamentalism is based on selective attention and equally selective inattention to religious texts.

The academic study of the Hindu and Buddhist traditions is as fraught with misconceptions as the environmental concerns of the last section. This is to some extent the legacy of colonial and missionary attitudes (though there have been notable exceptions), but it also reflects the determination of many westerners to 'discover' in Asia what they find lacking in their own societies. Many such people have great difficulty coming to terms with India's secular state, or the civic expression of south Asian Buddhism, based as it is on the Ashokan ideal rather than on the canonical texts. (And it is this that gives the tradition its social and environmental potential.) They therefore wonder why some monks seem uninterested in *nirvāṇa* and spend very little time meditating. Even more bewildering can be the discovery that the Thai word for the entity that survives death, *winyān*, is derived from one of the five Pali words used in orthodox Buddhism to *deny* the existence of any such thing (i.e. *viññāna*, the highest of the five components of personhood). The Thai don't even notice such inconsistencies. *Mâi' bhen rai'* . . .[4]

The literature dealing with religion and ecology cited in Chapter 1 is largely free from such misconceptions, but there is nonetheless a tendency in some scholarly works to superimpose an unrealistic agenda on south Asia. In *Purifying the Earthly Body of God*, for example, there is an article by Harold Coward on the notion of karma as a vehicle for promoting environmental awareness.[5] This is a perfectly sound and scholarly argument relating to a cardinal Hindu doctrine. However, we have already seen in Chapter 3 how Ram Mohan Roy jettisoned both karma and reincarnation in the name of rationalism, a view echoed by science students at several Indian universities I interviewed in the 1970s.[6] There is therefore little realistic prospect that karma will ever become a means of increasing ecological awareness among educated Hindus.

I hope that this study and the accompanying fieldwork will help to focus the work of future authors.

We shall review the previous chapters from the point of view of Stanley Tambiah's thesis that we must consider both the historical and the anthropological dimensions of religion in order to explore the continuities and transformations between them. As we do this, we must also expand our horizons – as Bridget and Raymond Allchin remind us – to take into account 'the ecological relationship between a human community or group and its environmental context'.[7]

Continuities

The earliest records of all the major religious traditions presuppose a close relationship between humanity and nature, and it is therefore quite easy to advocate environmental responsibility in terms of continuities between the past and the present by citing scriptural texts. I gave an example from the Judaeo-Christian tradition (St Luke's Gospel). Transformations are not so straightforward: my example was the transformation of the Hebrew notion of jubilee – the release of people and land every fifty years – as a basis for advancing the need to cancel the debts of developing countries to the international banks (which has considerable environmental implications for countries 'bonded' into growing cash crops for export). This was known as the Jubilee 2000 campaign, and the transformative religious symbolism gave it added momentum. In this section we shall review continuities which have environmental significance between the past and the present as we have encountered them in previous chapters; in the next section we shall consider the transformations.

The razing of the Khāṇḍava forest by Arjuna and Krishna (Chapter 2) does not in itself present any problem for environmentally conscious orthodox Hindus. As Karan Singh points out (Chapter 8) it may be interpreted in terms of the need to clear forests to provide for settled agriculture, but the manner in which they catch the escaping wildlife and throw it back into the flames, laughing and joking as they do so, is somewhat embarrassing. Admittedly, the Mahābhārata, from which this episode is taken, is an epic and not primary scripture (śruti), but Krishna is one of the most important divine incarnations and, together with Arjuna, occupies pride of place in the favourite of all Hindu sacred books, the Bhagavadgītā.

In Chapter 2 I offered a tentative account of some of the ways in which the early tradition must be enlarged in order to take ecological relationships into consideration. Thus the razing of the Khāṇḍava forest can be interpreted in terms of settled cultivators clearing land, and Agni 'burning along to the East' probably reflects the movement of Aryan settlers from the Punjab into the Gangetic valley. Much less certain is Indra's unblocking of the rivers interpreted (mainly by Kosambi) as the destruction of Indus Valley dams. However, we are probably correct in regarding Indra's powerful weapon (vajra) as the thunderbolt which releases the monsoon rains. More research is needed into the links between natural phenomena and scripture.

We also considered other aspects of the Hindu tradition which assign importance to the natural world. These include the sacred character of rivers and special places such as sacred groves and ponds, which are pre-Vedic in origin. In Chapter 8 M. C. Mehta described his belief in the sanctity of nature and his own sense of identity as a Hindu in terms of pre-Vedic traditions without reference to any literary texts. I summarized the four stages of the *varṇāśrama dharma* to stress their frugal and naturalistic character, and noted Śaṁkara's belief that the physical world is 'as real as we are'.[8] I emphasize this because some commentators who write about ecology and the Hindu tradition maintain that Śaṁkara taught that the world is unreal.[9]

I concluded Chapter 2 by summarizing the views of Karan Singh, an ecologically minded scholar whose exposition of Vedānta would find favour with many educated Hindus. According to Karan Singh, all that changes and moves is a manifestation of the eternal and unchanging reality that is *Brahman*, and all that lives is based on an undying *ātman*, which is our true or inner self. The external *Brahman* and the inner *ātman* are ultimately the same. The supreme goals of life are to recognize intuitively the existence of our true *ātman* within us, and to act in a morally positive manner for the welfare of all that exists.

Concern for the welfare of 'all that exists' has tended to be understood in terms of social responsibility. This was certainly the case in the nineteenth century, when many of the Hindu reformers tried to compete with the social improvements introduced by Christian missionaries. However, Karan Singh interprets welfare for all to bring out our responsibilities towards the natural world. In so doing he is enlarging the scope of our conventional anthropocentric world view to give it the ecological significance that it originally possessed. His Vedānta should therefore be seen as representing continuity with the past (in Tambiah's sense) rather than as a transformation.

Continuity with the past is also reflected in the writings of the more conservative nineteenth-century Hindu reformers, whose views are summarized in Chapter 3. Reasserters of religion such as Dayananda Sarasvati and the Arya Samaj were essentially fundamentalists who believed that all knowledge, including modern science, is to be found in the Vedas. In recent years this point of view has tended to be represented by the publications of the Bharatiya Vidya Bhavan, especially *Bhavan's Journal*, which functions very much as the *Reader's Digest* of urban Hindu India.[10] This and similar publications in Hindi and the regional languages discuss the potential of religion for environmental improvement. The Hindu Right has yet to propose a convincing environmental ethic; we noted some possibilities in Chapter 7.

Turning from the Hindu to the Buddhist tradition, we recall that the Buddha's teachings were full of allusions to the natural world. He denied the Hindu notion of a soul, describing personhood in terms of five aggregates. Non-human living beings were inhabited by spirits, and the earliest monastic rules contain injunctions against the destruction of even the most primitive life forms. It is the Buddha's fundamental message, however, with its clinical analysis of the roots of human suffering in terms of acquisitiveness and individualism, that represents the greatest challenge to the forces that are responsible for so many of our current social and environmental ills. The Four Noble Truths and the Noble Eightfold Path, though anthropocentric in focus, set out the manner in which we are going wrong in the way we live, and tell us how to do better. They represent a direct continuity between the past and the present.

Early Buddhism did not follow the Vedic custom of sacrificing animals. It appeared later than the Hindu tradition and well after the completion of the three stages of resource use that I described in Chapter 2. It was only when Ashoka, building on a common stock of Indo-Aryan politico-moral ideas, set himself up as *dharmarāja* and *cakravartin* and proclaimed his conservationist edicts, that Buddhism began to acquire the ecological thrust which subsequently characterized the Theravāda in south and southeast Asia.

Walpola Rahula and others have argued that when Buddhism was adopted by Ashoka it ceased to be a religion and developed into an ecclesiastical organization whose preoccupation with social and political matters hampered its vitality.[11] By contrast, I maintain that Buddhism began as a theory of human existence with implications for human society to which Ashoka and his successors ultimately gave concrete expression. According to Trevor Ling:

> The Buddha was an 'analyst', not a propounder of dogmatic truth, and early Buddhism was characterized essentially by its rationalism . . . The human 'need' to which the Buddha addressed himself was not that of man's need for religion, but man's need to overcome his condition of self-centredness, and to identify with a greater, completely comprehensive reality. If man has any innate spiritual 'need', it would appear to be this, rather than religion.
> . . . In some Asian countries Buddhism retains a good deal of its original concern with the public dimension of life as distinct from the private world of soul-salvation, its character as an ideology capable of integrating a religiously and even culturally pluralistic society. It is in Western countries, on the whole, that there is the strongest insistence on regarding it as a religion competing with other religions.[12]

Such a view is consistent with what I have argued in Chapters 5 and 6, namely that it is the social and civic dimensions of Buddhism, which can be traced back to Ashoka, which have provided the thrust for the *saṅgha*'s current social and environmental concern in Theravāda countries, especially Thailand. I offered an account of the contemporary roles played by Thai monks, whose outlook has been strongly influenced by the scriptural conservatism of Mongkut's ecclesiastical reforms. Many monks continue to excel as scholars and meditation teachers, though an increasing number are taking up development-oriented activities which include a significant environmental component. The former group represents continuity with the past as reflected by canonical scripture; the latter are the imaginative transformers of it.

Transformations

All transformations of tradition share an element of continuity with it. Thus Ashoka's model of the monarchy, though based on his distinctive self-understanding, also represents continuity with Hindu and Buddhist notions of kingship (I have avoided detailed discussion of these by referring to a common stock of Indo-Aryan ideas). However, when Ashoka assumes to himself cosmic roles such as we noted in Chapter 5, he is transforming the past in an imaginative manner.

Transformations of the past are much more likely to cut across the various dimensions of religion outlined by Ninian Smart in Chapter 1. For this reason they are much less likely to appeal to religious fundamentalists, who consider scripture to be the ultimate authority on all matters of belief and conduct. It is the transformations that open up new possibilities, however, and this is especially true in relation to social and environmental concerns.

The nineteenth-century reforms in India paved the way for more recent attempts to address environmental problems in the name of religion, but the reformers were primarily interested in social rather than environmental issues. They were unaware of the full consequences of European industrialization and of the extractive mode of resource use which was to erode the natural resource base of first south and later southeast Asia.

Of the various reformers, Vivekananda was the most influential, and we have seen in Chapter 3 how his view of karma-*yoga*, this-worldly action, paved the way for Gandhian ethics. However, at the point where his concern for 'the lowest worm that crawls under our feet' might have moved him away from anthropocentrism, he advocated social action for the sake of others: 'Why should I love everyone?' he demands. 'Because they and I are one . . . There is this oneness, this solidarity of the whole universe. From the lowest worm that crawls under our feet . . .'

Such sentiments are consistent with *advaita* and represent continuity with tradition, but as an expression of our human responsibilities, they tip the balance of karma-*yoga* very much in favour of what we do here and now rather than in terms of past or future lives. The concept of personal and collective service for others in terms of the kind of educational and medical programmes undertaken by the Ramakrishna Mission, though influenced by both Islam and the activities of Christian missionaries, represents a significant transformation of the Hindu tradition.

Gandhi's belief in 'the essential unity . . . of all that lives' builds on Vivekananda's ideas and forms the basis for his cosmocentric anthropology. It represents continuity with the past. However, *ahiṁsā* (soul force or insistence on the truth), *satyāgraha* (laying hold of the truth or reality), *swaraj* (participatory democracy), *swadeshi* (identification with one's own country) and *sarvodaya* (the awakening of all) are all imaginative transformations of tradition with considerable implications for ecology. Thus *sarvodaya*, for example, expressed in practical terms as the establishment of participative village republics, implies a long-term sustainable relationship between people and their natural environment. *Satyāgraha*, which provides a suitable vehicle for the practice of *ahiṁsā*, is the expression of love for all, even 'the meanest of creation'. These are all transformations of the Hindu tradition, though it is worth noting in passing that in Sri Lanka, A. T. Ariyaratne has redesignated them as Buddhist.

Gandhi's contemporary, the Bengali scientist Jagadish Chandra Bose, may at first sight seem an unusual addition to my list of ecologically significant Hindu reformers, but a number of contemporary environmentalists, including Sunderlal Bahuguna, pay tribute to his influence. In many ways J. C. Bose achieved the most imaginative of all transformative links between the past and the present. Noting the coming together of the various branches of science during the greater part of the nineteenth century, he concluded that this was evidence of a convergence in which the many manifestations of reality were becoming one. He therefore decided to conduct his experiments in the border regions between various scientific disciplines, researching such unconventional topics as the possibility of pain in plants. Such a transformation of Vedānta to provide an imaginative agenda for empirical research was a remarkable achievement. I shall say more about science presently.

The Chipko environmentalists and Anna Hazare were Gandhian in their outlook and activities (Chapter 4), but Sunderlal Bahuguna's *padayātrās* along the length of the Himalayas are part of a much older tradition of pilgrimage based on the belief that every landscape is imbued with a numinous and spiritually transformative power. David Kinsley describes this:

> Pilgrimage is often the process of learning to see the underlying or implicit spiritual structure of the land; this often involves a change in perspective, a change that is religiously transformative. Pilgrimage is the process whereby pilgrims open themselves to the sacred power, the numinous quality, of the landscape, whereby they establish a rapport with the land that is spiritually empowering. An underlying assumption of pilgrimage seems to be that the land cannot be intensely known and experienced from a distance; it can be fully known, its story deeply appreciated, only by travelling the land itself. The physical immediacy of pilgrimage, the actual contact with the land, intensifies the experience of appropriating the story of the land, learning to see its underlying, implicit structure, sensing its spiritually enlivening power. The experience can be lasting, transforming one's perspective permanently.[13]

If all land is sacred, then the territory along the perimeter of the Himalayas, the holy mountains where trees, rivers and the perennial snows are the home of the gods, must be even more so. In this context Sunderlal Bahuguna's *padayātrās* have become powerful transformative symbols.

Turning to Buddhism, we have seen how Ashoka's rule in India mediated some of the most important transformations of the Theravāda tradition as it moved to Sri Lanka and, later, southeast Asia. His understanding of his rule extended beyond human affairs to include the entire cosmos; on a more day-to-day basis he went to great lengths to conserve forests and protect their denizens.

Mahāyāna Buddhism took root in the north of India in the Himalayan regions and beyond, where it incorporated elements such as the *bodhisattva*, whose compassion extends to the whole of creation, Tibetan Bön belief in nature worship and spirits, and the practical and this-world immediacy of *tantra*. We saw in Chapter 5 how the Buddhism of Ladakh and Bhutan enabled monks, nuns and lay Buddhists to take part in activities to promote social and environmental improvement. Monasteries can be used, for example, to demonstrate appropriate technology and renewable energy sources. For all that Ladakh and Bhutan both face serious social and political problems, they may nonetheless be said to demonstrate what Donald Swearer describes in another context as 'compatibility between the Buddhist world view of interdependence and an "environmentally-friendly" way of living in the world'.[14] Buddhism's traditional interdependence between human life and the natural world has been transformed to give it considerable contemporary relevance.

In Buddhism actions and their consequences are determined by *paṭicca samuppāda*, or interdependent co-arising, which embraces all life and non-life in a continuous web of interdependence. Buddhadāsa's views in Chapter 6 that only *śūnyatā* truly exists, that *nibbāna* is here and now, and that our wholesome actions move us away from ego-centredness in accordance with a reinterpreted doctrine of no-soul, represent a transformation of traditional Buddhism which has major implications for our responsibilities for the natural world. According to Donald Swearer, these can be elaborated in terms of our need to conserve nature (*anurak thamachāt*) with a quality of empathetic care that lies at the core of our being, so much so, in fact, that 'I' cease to exist over and against nature (with all my self-centred individualism). Instead, I become interactively codependent with it.

The extent to which the Hindu and Buddhist traditions encourage women to participate actively in social and environmental improvement has not been discussed in detail, except in the case of Thailand (Chapter 6), where I identified the lay nun (*mae chii*) as having a potentially significant role. Far from being poor and elderly widows who hide themselves at the backs of temples, modern *mae chii* are redefining their roles in socially and pastorally relevant ways which complement, and in some cases improve on, the monks' more narrowly and scripturally defined responsibilities. This can be represented as the transformation of tradition, because *mae chii* have historical precedents, though not in relation to scripture, because the Pali canon makes no mention of them, and in any case the Buddha's views about the role of women are unclear.

My arguments are supported by fieldwork I have conducted. Thus Appendix A lists a variety of medicinal plants identified at Buddhist temples in Thailand, the locations of which are shown in Figure 6.1 (p. 87). The cultivation of such medicinal plants is advocated in both Buddhist and Hindu religious texts, and it only began to decline with the advent of western medicine. In the light of recent concern to preserve biodiversity, however, the monks' inherited knowledge not only represents a transformation of the wisdom of the past, but is a vital resource which must be encouraged through the kind of educational programmes pioneered by Prawase Wasi, who would almost certainly have received a Nobel Prize had he conducted his research more prominently in the West.

I have not discussed Islam in south and mainland southeast Asia, a vast topic which merits a separate study. Malaysia is home to a considerable amount of social and environmental religious concern, and several of India's Muslim organizations are aware of the need to conserve resources. Kashmiri Muslims deplore the eutrophication of the Dal Lake, for example, and the main mosque in Srinagar uses solar energy to supply electricity. It is also significant that the World Muslim Congress, based in Karachi, took part in the Earth Summit.

Christian interest in social and environmental issues is also worthy of further research. All the major south Asian churches have an impressive record of involvement in social issues, though this has been under threat in recent years as a result of the activities of US-based 'Christian' fundamentalist organizations. Many Roman Catholic schools and colleges in India take a lively interest in environmental issues, and we have seen in Chapter 8 how Thomas Kocherry, a priest in Kerala, was able to protect the rights of fishing communities against coastal violations by foreign ships.

Public-interest science

We noted in the last section that Jagadish Chandra Bose was able to transform the Hindu tradition so imaginatively as to use its philosophical presuppositions to guide his scientific research. The quality of his work has been universally recognized; it paves the way for approaches to science which are much more environmentally sensitive and culturally appropriate than the ones most people are familiar with.

Science and technology play a considerable part in ecological relationships between people and the natural world. They are often asserted to be value-free, but this is seldom the case. For example, if science operates according to the well-established principles of scientific forestry, then it will systematically replace existing broad-leaved species of tree, such as oak, with monocultures of, say, *chir* pine, which can be selectively or clear felled for commercial profit. The disastrous long-term ecological consequences of such policies have been described.

The *modus operandi* of science in such situations has been summarized by Vandana Shiva:

> Scientific knowledge is not [as] universal, objective and neutral as it is posited to be. It is always a particular response to a particular interest. When the interest is the commercial utilization of resources for maximizing exchange value, the type of knowledge system that is created is reductionist. Internalization of profits and externalization of costs is a normal consequence when nature is treated as if its individual components are isolated and unrelated, and the only components with economic value are those that can be transformed into commodities. The basic terms, concepts and definitions have built into them the economic values of the interest to which the knowledge is a response.[15]

When the interest is broader and more comprehensive, including such factors as the sustainable livelihood of communities living in or near a forest and the satisfaction of basic human needs, then a different type of science is required. At the very least such public interest science will include the following:

1 The satisfaction of basic human needs.
2 Economic development with sustainability.
3 The equitable distribution of the costs of development.

The implications for science are that it will need to learn much more about ecological factors that are often best understood by the people who live closest to nature: farmers, fisherfolk, tribals, monks, etc. Secondly, science must be willing to operate within a much more interdisciplinary framework than hitherto. One of the main reasons why western science has become problematic is that its terms of reference are often determined in relation to a single discipline (e.g. economics or physics), which does not take into account the implications for other sectors. This is not merely an issue which ecologically minded enthusiasts raise in order to justify their claims; even a hard-nosed physicist writing recently in a professional journal complains about what he calls the pervasive spread of econometric forms of quantification:

> The assignment of monetary value to each and every facet of existence and to every possible transaction represents a pathological takeover by quantitative economics of just about everything. People are being led to believe that everything in society has an assignable cash value; anything that does not is fading from sight. A generation is growing up that believes that rigorous costing is a feature of nature and that everything has a 'real' cost – just as physical objects have a 'real' rest mass or momentum. In truth, cost assignments are possible only if the calculations are applied in strictly limited contexts.[16]

Reductionist science and unbridled monetarism are major contributors to the consumerism which I deplored in Chapter 1.

Culture and religion can play an enormous part in broadening the scope of science and in making it more sensitive to its ecological context. Much of the debate that has occurred between science and religion has assumed science to be value- and culture-free, when in reality its context has been entirely western. Western science has then been related, without too much difficulty, to western religion.

Jagadish Chandra Bose and some of his colleagues broke away from the western mould of science to explore the frontiers between different scientific disciplines. He earned the respect of the international scientific community and was also able to influence the beliefs of Sunderlal Bahuguna and other contemporary environmentalists.

In 1896 Lord Hamilton wrote about Bose to Lord Elgin as follows:

> There is a strong feeling here that the Government should in some way mark its appreciation of Dr J. C. Bose's remarkable labour and researches in science. The highest scientists here express great admiration of the little man . . .[17]

May we hope that this 'little man' will one day be acknowledged as the pioneer of new forms of environmentally sensitive and culturally appropriate science?

Environmentalists at the crossroads

The United Nations report on environment and development which paved the way for the Earth Summit was extremely optimistic about the role of science and technology in developing countries:

> Blends of traditional and modern technologies offer possibilities for improving nutrition and increasing rural employment on a sustainable basis. Biotechnology, including tissue culture techniques, technologies for preparing value-added products from biomass, microelectronics, computer sciences, satellite imagery and communication technology are all aspects of frontier technologies that can improve agricultural productivity and resource management.[18]

It is therefore very unfortunate that so many environmentalists, especially in the West, have come to regard science and technology as harmful. Contributors to *Buddhism and Ecology*, for example, cited factually incorrect information to claim that the safe disposal of radioactive nuclear waste is impossible and, by implication, that nuclear power should be abandoned.[19] Environmental organizations which once helped to curb the environmental recklessness of large corporations now raise public fears quite unnecessarily about the safety of genetically modified crops which, subject to careful testing, could be of immense benefit to developing economies.

Indian environmentalists have so far refused to turn their backs on science and technology. Within the Chipko movement, Chandi Prasad Bhatt has pioneered appropriate technology (Chapter 4). The Marxist Uttarakhand Sangharsh Vahini, in common with most Marxist groups, is optimistic about science and technology, and Sunderlal Bahuguna, though currently against the building of large dams, is not opposed to science; this is clear from his respect for Jagadish Chandra Bose. Mahatma Gandhi, whose views shaped those of many Indian environmentalists, was not against science and technology *per se*, though he inveighed against the excesses of industrialization.

Many of the Indian NGOs which participated in the Earth Summit train professionals – scientists, engineers and lawyers – in environmental awareness. Many more advocate appropriate technology, often drawing upon the wisdom of the past (e.g. water harvesting). The environmental activities that I reviewed in Chapter 8 include the extensive use of appropriate technology such as micro-hydro, check dams and sophisticated tissue culture techniques. The extent to which some Indian environmentalists have distanced themselves from their western counterparts may be illustrated by Thomas Kocherry, who refused to accept a prize for which he was nominated by

Greenpeace because the prize money had been donated by an oil company whose ships polluted coastal waters.

In Thailand, environmental activists, such as Sulak Sivaraksa, Phra Prajak and other environmentalist monks, oppose many technological developments because they will benefit only the richest members of society and fuel the consumerist forces which feed into global markets. The Thai, like Indians and members of other south and southeast Asian societies, are proud of their scientific and technological achievements, and encourage their young to obtain qualifications and pursue careers in these areas.

At the international level, environmental organizations constitute one of three main types of NGO which mushroomed during the 1990s, the others being in the fields of development and human rights. The World Wide Fund for Nature has increased its membership tenfold, to about five million, since the mid-1980s. Friends of the Earth has one million members in fifty-eight countries. Greenpeace has two-and-a-half million members and 1,100 staff. Together they represent an enormous worldwide potential for social and environmental improvement.

Environmental NGOs are often more effective than governments in responding to environmental crises. They can mobilize grass-roots support and public opinion through the media, and with the global economy imminent, they are well placed to challenge the excesses of transnationals and irresponsible governments. Thus Greenpeace performed an important service to the international community by opposing French nuclear weapons testing in the Pacific, a move which led to the blowing up of its ship, *Rainbow Warrior*, at the instigation of French government agents. It was justified in opposing big business monopolies over the production and distribution of genetically modified crops, and may even have been right in April 2000 to challenge an attempt to drill for oil in the Arctic.

However, there are major divisions among the various international 'green' organizations over policy matters, and several of them display considerable ruthlessness towards the others when they fall out. In 1993, for example, Greenpeace published a book of all its environmental rivals, some of which differed so insignificantly from them as to suggest that the real issue was that they were all competing for the same funds. The presence of 9,000 journalists at the Earth Summit in Rio de Janeiro in 1992 created a climate of public opinion in which several NGOs felt constrained to pursue popular goals which were not in line with the expectations of their donors.

Patrick Moore, a founding member of Greenpeace, claims in a recent *New Scientist* interview that the international environmental movement began to lose its sense of direction in the mid-1980s:

> I'm a Gandhian through and through – I believe that civil disobedience and passive resistance movements are great shapers of social change. But when industry and government agree that the environment needs to be taken into account in policy making, . . . it seems to me it would be a

good idea to work with them . . . The environmental movement abandoned science and logic somewhere in the mid-1980s, just as mainstream society was adopting all the more reasonable items on the environmental agenda.

Environmentalism has become codified to such an extent that if you disagree with a single word, then you are apparently not an environmentalist. Rational discord is being discouraged. It has too many of the hallmarks of the Hitler youth, or the religious right.[20]

According to Richard Jefferson, a molecular biologist who helps farmers in developing countries:

In their anti-corporatism [the environmentalists] have become anti-technology, anti-science, anti-informed discussion. They do not like complexity, just the black and white.[21]

More moderate environmental organizations include the International Consultancy on Religion, Education and Culture (ICOREC), which has done excellent work on raising environmental awareness on an interfaith basis.[22]

Secular India

The fact that Buddhism began both as a view of human existence with implications for human society and as a philosophy with no need of theistic beliefs or sanctions yet tolerant of them makes the Buddhist sculpture known as the Sarnath Lion Capital of Ashoka an extremely appropriate emblem for the modern, secular Indian state. In Chapter 7 we considered the main features of the Indian secular state and attempts by various governments since Independence to implement its ideals.

The policies of successive Indian governments to eradicate poverty were thwarted by tensions between the centre and the states, caste alliances and a variety of other factors. Five-year plans foundered for reasons to do with over-reliance on heavy and basic industries, which in turn exacted a high environmental cost. The 'green revolution' increased the gap between rich and poor and caused further environmental problems, which culminated in the Bhopal disaster in 1984. Nehru's vision of a new India dovetailed all too neatly into the less altruistic ambitions of urban and rural élites and paved the way for the thoroughgoing exploitation of natural resources to feed the international markets. Mahatma Gandhi's alternative model of crafting an agrarian society of village republics making low levels of demand on natural resources by living close to subsistence was bypassed. We also considered the potential of the Hindu Right to address social and environmental problems.

Prior to and following the 1992 Earth Summit the Indian Government became involved in international discussions on ozone depletion, climate change and biodiversity loss. I described these discussions against the background of the more immediate needs of ordinary people as reflected by the concerns of the Indian NGOs at the Earth Summit (Appendix B). I summarized these as deforestation, waterway and other forms of pollution, healthcare, women's education, resource depletion and the basic needs of predominantly rural communities. Local solutions to these problems are to be found in the eco-development programme of Anna Hazare (Chapter 4) and the various initiatives described in Chapter 8. The activities of the more notable environmentalists are beneficial in terms of raising public awareness, but controversial in other respects.

Gandhi once made the observation that to a hungry person, not even God dare appear except in the form of food. We believe that the preservation of human life must take precedence over environmental conservation if and when such a choice has to be made, but the best development will always go hand in hand with the enhancement of the natural world as measured by public-interest science. Ecology is a branch of science that must be taken much more seriously than at present, especially in the West, where genuine science is increasingly being replaced by technological reductionism.

In Chapter 7 I summarized the analyses of India's environment and development problems by two pairs of authors, Madhav Gadgil and Ramachandra Guha, and Jean Drèze and Amartya Sen. The first pair use an environment-led approach to adduce six root causes of India's problems, ranging from the narrowness of the natural resource base available to most people, to heavy dependence on imported technology and petroleum products. They proceed to discuss the extent to which the political philosophies which emerged from the Chipko and Appiko movements are comprehensive enough to address these root causes, concluding that each has strengths and weaknesses. They argue that a combination of components of Gandhism, Marxism and liberal capitalism points in the direction of nine practical measures; these include a form of participatory democracy in which both the political leadership and the bureaucracy are accountable to ordinary people, and new strategies for agriculture and industry requiring lower levels of energy input.

Jean Drèze and Amartya Sen use a development-led approach to argue the case for a broad and participative interpretation of economic development which takes into account the need to expand social opportunities. Where social opportunities exist, these authors maintain that the enlargement of markets can enhance them. However, people are excluded from benefits through illiteracy and lack of education, together with other capabilities related to basic health, gender differences, social security, land rights and local democracy. The need is for strong initiatives by central government – the secular state – accompanied by substantial nationwide participatory programmes of primary and secondary education, healthcare and the empowerment of women.

While acknowledging the value of Gadgil and Guha's analysis, my preference is for the arguments of Drèze and Sen, on the grounds that the latter offer a more comprehensive and realistic context for many of the former's recommendations, and that their views are applicable to other developing countries (e.g. in southeast Asia). Their emphasis on local participation is also compatible with education designed to promote environmental awareness and the responsible use of natural resources.

Chapter 8 contains a series of interviews with the kind of secular, educated Hindus that one might meet at an international gathering or during a visit to a major Indian city. They are not representative of India as a whole, but they are influential in many spheres of public life. Their views are extremely diverse. Thus Karan Singh, representing an orthodox Hindu position, advocates the incorporation of environmental values into formal and non-formal education and such practical measures as the restoration of the festival of trees (*vana mahotsava*). M. C. Mehta disregards the Hindu religious texts, and harks back to the pre-Vedic attribution of sanctity to rivers and trees. Thus the custom of touching a baby's lips with water from a sacred river (Ganga *jal*) could become a reminder of the importance of water and, insofar as the practice is not specifically Hindu, might become equally acceptable to Muslims and Christians.

Amartya Sen's analysis of the Bhagavadgītā in Chapter 8 represents a much more thoroughgoing critique of Hindu orthodoxy than M. C. Mehta's views. His understanding of consequentialist ethics in terms of taking responsibility for one's actions is as foundational to his interpretation of Arjuna's predicament *vis-à-vis* the killing of his own kinspeople as it is to his views on the alleviation of poverty (which is concerned with deprivations which include so-called ecological deprivations). Such an interpretation of a cardinal Hindu text runs counter to prevailing Hindu orthodoxy, but lifts the relevance of the dialogue onto a contemporary level. Whether we are dealing with ecology or Arjuna's predicament, we must learn to take responsibility for all our actions. Such a consequentialist approach is a much bolder transformation of religion – all religion – than any of its more conventional expositions.

I hope I have answered the questions posed at the end of the introductory chapter as to why there was so much Indian participation at the Earth Summit and why the Indian and Thai governments felt able to present their cases to some extent in religious terms. I hope that my broad view of south and southeast Asia will help to focus further research into ways in which the Hindu and Buddhist traditions can increase ecological awareness and contribute to the resolution of social and environmental problems that will affect the fabric of our planet and the quality of life on it well into the new millennium.

Appendix A
Medicinal plants identified in Thailand

The following medicinal plants were identified in the precincts of Buddhist temples at locations indicated on the map of Thailand (Figure 6.1). Monks supplied the Thai names for the plants and described their use. I photographed the plants and their further characteristics as listed were obtained from a botanical catalogue at Chulalongkorn University. For further details see notes 21 and 22 in Chapter 6.

Thai	English	Latin	Family	Parts used	Uses
Borapet	Cactus	Tinospora tuberculata	Menispermaceae	Stem	Fever
Eucalyp	Blue Gum	Eucalyptus globulus Labill	Myrtaceae	Leaf	Antiseptic, anti-flatulent
Fang	Sappan	Caesalpinia sappan Linn.	Caesalpiniaceae	Wooden stem	Menstrual problems, diarrhoea
Fin Ton	Coral Plant	Japtropha multifida	Euphorbiaceae	Latex from stem	Laxative
Hanumankopsap					Toxification
Hanuman Nangten					Wounds
Hanuman Prasanguy		Schefflera venulosa Harms	Araliaceae	Young leaf	Asthma, haemostasis
Harroynang					Toxification
Keelek	Siamese Cassia	Cassia siamea Lamk.	Caesalpiniaceae	Flower, young leaf	Laxative, appetizer, sedative
Keetun					Gonorrhoea
Khae (Khae Khaw, Khae Daeng)	Cork Wood	Sesbania grandiflora Pers.	Papilionaceae	Bark	Diarrhoea
Kha Jurt					Travel sickness
Khamin Khaw					Menstrual disorders, blood disease
Khamin Oi	Zedoray	Curcuma zedoaria Rosc.	Zingiberaceae	Rhizome (root)	Anti-flatulent
Khem Daeng		Ixora coccinea Linn.	Rubiaceae	Root, flower	Fever, diarrhoea

Thai	English	Latin	Family	Parts used	Uses
Khem Khaw		Ghasalia curviflora	Rubiaceae		Gonorrhoea
Khing	Ginger	Zingiber officinale Rosc.	Zingiberaceae	Rhizome	Carminative, skin problems
Khinghengplakang					Menstrual disorders
Khlu		Pluchia Indica (L.) Less.	Compositae	Whole plant, bark, leaf, root, sap	Diuretic, sinus disorders
Kiajiab Priaw (Kiajiab Daeng)	Red Sorrel (Roxella)	Hibiscus sabdariffa Linn.	Malvaceae	Calyx	Hypertension, urinary bladder stones
Krachai Daeng					Menstrual disorders, blood disease
Ling Dam					Toxification
Maduachumporn		Ficus glomeratus	Moraceae	Root	Fever
Mahajakaphart					Diarrhoea
Mahamek (Wan Mahamek)		Curcuma aeruginosa	Zingiberaceae		Diarrhoea
Matum (Mapin)	Elephant's Apple	Aegle marmelos Coor.	Rutaceae	Young fruit, leaf, ripe fruit	Diarrhoea, chest problems
Naam Klet					Kidney disease
Nguaplaamor					Haemorrhoids
Noinaa	Sweet Apple, Custard Apple	Annona squamosa Linn.	Annonaceae	Seed, unripe fruit, root	Purgative, skin diseases, lice

Thai	English	Latin	Family	Parts used	Uses
Nok Hongyok					Gonorrhoea
Pengpuay Falang		Catharanthus roseus			Leukaemia
Petsangkhat (Sam Roi Tor, Khan Khor)		Cissus quadrangularis Linn.	Vitaceae	Sap, root, young leaf	Removing 'yellow blood' from ear, nosebleeds, dressing wounds and broken bones, menstrual problems
Phak Kachet		Neptunia oleracea Lour.	Mimosaceae	Whole plant	Toxification (stomach)
Phayaa Sataban (Tin Pet Jet Ngam)	Dita or White Cheese Wood	Alstonia scholaris (L.) R. Br.	Apocynaceae	Bark, latex, young leaf, seed	Fever, malaria, diarrhoea, stomach 'heat', menstrual disorders
Phlap Phlung (or Phlap Phlung Dok Daeng, Phlap Phlung Salap Khaw)		Crinum amabile Dorn.	Amaryllidaceae	Bulb, leaf	Emetic (bulb), joint pains (leaf)
Phrai Dam		Zingiber ottensii Valeton	Zingiberaceae	Rhizome, leaf	Diarrhoea, sedative
Phyagarsak					Diabetes
Sabu Luat (Hang Yai Khlayhin)					Toxification
Sabu Luat (Thaw)					Toxification

Thai	English	Latin	Family	Parts used	Uses
Sadow India	Neem Tree (Margosa)	Azadirachta Indica (L.) Juss.	Meliaceae	Bark, leaf, seed	Diarrhoea, prevention of malaria
Salao (Salao Khaw)	Lagerstroemia	Tomentosa Presl.	Lythraceae	Bark	Diarrhoea
Salchpangporn		L. clinacanthus Nutans	Acanthaceae	Leaf	Animal bites
Tam Lung Tua Phu		Melothria heterophylla	Cucurbitaceae		Skin rash
Wan Garb Hoy (Wan Hoy Khreng)	Boat Lily	Rhoeo discolor Hanse	Commelinaceae	Leaf	Sore throat, cough, internal haemorrhage, anaemia
Wan Singhamora (Phak Nam Farang)		Cyrtosperma. johnsonii N.E. Br	Araceae	Leaf, flower, whole plant, rhizome	Asthma, menstrual problems, indigestion, headache
Warn Nam	Sweet Flag	Acorus calamus Linn.	Araceae		Stroke
Ya Khiaw (Rangchut)	Blue Thunbergia or Laurel Clockvine	Thunbergia laurifolia Lindl.	Thunbergiaceae	Leaf	Fever, wounds

Appendix B
Indian non-governmental organizations (NGOs) accredited by the United Nations which participated in the Earth Summit in Rio de Janeiro in 1992

1. **Academy of Gandhian Studies**, Hyderabad, 1976: promotes educational and practical issues (e.g. organic farming) within the context of Gandhian philosophy.
2. **Action Association for Rural and Tribal Development**, Jangareddigudem (Andhra Pradesh), 1988: organizes relief and reconstruction programmes for the victims of natural calamities (e.g. floods and cyclones).
3. **All India Women's Conference**, Delhi, 1927: promotes women's education through camps and seminars in tree planting and water conservation.
4. **All India Women's Studies and Development Organization**, Kanpur (Uttar Pradesh), 1989: promotes the education of women in family planning and self-help.
5. **Asian Environmental Society**, Delhi, 1971: promotes public awareness of environmental issues and sustainable development.
6. **Association for the Propagation of Indigenous Genetic Resources**, Ahmedabad (Gujarat), 1987: supports the use of indigenous genetic resources and opposes chemical fertilizers and pesticides.
7. **Association of Hydrologists of India**, Visakhapatnam (Andhra Pradesh), 1981: organizes training courses for professionals on water conservation.
8. **Baif Development Research Foundation**, Pune, 1967: organizes agro-based rural development programmes and environmental education for children.
9. **Balmunch**, Ujjain (Madhya Pradesh), 1985: promotes cultural programmes on global warming and energy conservation and trekking to raise environmental awareness.
10. **Bangalore Integrated Rural Development Society**, Bangalore, 1977: organizes education programmes on environmental protection and vocational training courses for women.
11. **Bombay Environmental Action Group,** Mumbai, 1979: campaigns

against polluting urban industries and initiates legal action to protect forests.
12 **Bombay Natural History Society**, Mumbai, 1883: promotes the conservation of wildlife.
13 **Centre for Culture and Development**, Chennai, 1990: promotes culturally appropriate participatory development and indigenous health practices.
14 **Centre for Environment Education**, Ahmedabad (Gujarat), 1984: supports media campaigns to raise environmental awareness among rural communities and to encourage collaboration between NGOs and the Government on environmental issues (e.g. joint forest management).
15 **Centre for Environmental and Management Studies**, Delhi, 1986: promotes seminars for industrialists and educationalists on environmental issues.
16 **Centre for Mass Communication**, Chennai, 1985: collaborates with the government in communicating global environmental issues via regional languages in south India.
17 **Centre for Research on Sustainable Agricultural and Rural Development**, Chennai, 1990: promotes sustainable and equitable rural and marine development.
18 **Centre for Research, Planning and Action**, Delhi (details not available).
19 **Centre for Science and Environment**, Delhi, 1980: aims to increase public awareness of environmental issues in India and globally through workshops and publications (e.g. *Down to Earth*).
20 **Centre for Science and Technology of the Non-aligned and Other Developing Countries**, Delhi, 1989: promotes the use of science and technology at regional and national centres for development (e.g. low-cost housing, biotechnology and renewable energy).
21 **Centre for the Study of Man and the Environment**, Calcutta, 1978: promotes the involvement of professional scientists in environmental issues through research projects and education.
22 **Chemtech Foundation**, Mumbai, 1970: advises the government on air and water quality, solid waste and industrial safety.
23 **Community Service Centre**, Chennai, 1968: organizes educational programmes related to church and society on social and ecological issues.
24 **The Community Services Guild**, Chennai, 1980: organizes educational programmes on afforestation, agriculture and sanitation.
25 **Consumer Education and Research Centre**, Ahmedabad (Gujarat), 1978: promotes public awareness in industrial safety.
26 **Consumer Education Centre**, Hyderabad, 1982: publishes a quarterly magazine called *Consumer Network News of India* which deals with environmental issues as they relate to consumers.
27 **Consumer Unity and Trust Society**, Calcutta, 1980: publishes material and organizes workshops on links between environmental degradation and poverty.
28 **Council of Professional Social Workers**, Bhubaneswar (Orissa), 1986:

organizes meetings on industrial pollution, women's issues and the rights of indigenous people.
29 **C.P.R. Environmental Education Centre**, Chennai, 1988: promotes programmes on water resources management, wasteland development and family planning.
30 **Deccan Development Society**, Hyderabad, 1983: promotes activities dealing with women's issues, human rights and education for the children of the rural poor.
31 **Development Alternatives,** Delhi, foundation date not specified: involved in national follow-up of the Earth Summit.
32 **Elders Council of India**, Hyderabad, 1983: conducts seminars to encourage professionals to create a healthy environment.
33 **Energy Environment Group**, Delhi, 1985: promotes energy-efficient agriculture and technology through publications and media coverage; translated *Our Common Future* into Hindi.
34 **Environment Preservation Society**, Gondal (state not specified), 1979: organizes nature conservation activities.
35 **Environment Society of Chandigarh**, Chandigarh, 1976: promotes educational activities on environment and development issues in schools and colleges.
36 **Evangelical Fellowship of India Commission on Relief**, Bangalore, 1967: organizes activities to assist local communities in watershed management, income generation and women's issues.
37 **Federation of Voluntary Organizations for Rural Development in Karnataka**, Bangalore, 1983: acts as umbrella organization for rural development agencies.
38 **Global Futures Network**, Mumbai, 1981: promotes long-term planning for a sustainable future via publications on the environment, science and technology.
39 **Gokul Prakalp Pratishthan**, Ratnagiri (Maharashtra), 1978: promotes bioregional networks and ecologically planned villages in which people can choose occupations and duties according to skill and age.
40 **Gondwana Geological Society**, Nagpur (Madhya Pradesh), 1981: organizes environmental meetings from a geological standpoint.
41 **Gorakhpur Environmental Action Group**, Gorakhpur (Uttar Pradesh), foundation date not specified: attempts to provide solutions to problems of acid rain, global warming and ozone-layer depletion.
42 **Gram Swarajaya Samity Bakhtuyarpur**, Patna (Bihar), 1978: operates on two fronts, one practical and geared to such issues as improving air quality by changing from coal to biogas, the other a spiritual awareness programme.
43 **Gramodaya** (Village Awakening), Manapparai (Tamil Nadu), 1982: promotes the integration of local knowledge into self-help schemes in afforestation, waste land development, rural sanitation and alternative technology.

44 **Indian Agrometeorological Society**, Visakhapatnam (Andhra Pradesh), 1982: collaborates closely with the Asian Climatological Society in studies of desertification, environmental degradation and floods.
45 **Indian Association for the Advancement of Science**, Delhi, 1977: involved in national survey of different agro-climatic zones to identify training needs for sustainable agriculture.
46 **Indian Committee of Youth Organizations**, Delhi, 1983: coordinates youth environmental activities.
47 **Indian Community Development Service Society**, Bangalore, 1983: runs a model community irrigation system for poor farmers as part of a programme to evolve alternative social structures.
48 **Indian Institute of Forest Management**, Bhopal (Madhya Pradesh) 1982: promotes the conservation of forests within a national framework.
49 **Indian Institute of Youth and Development**, Phulbani (Orissa), 1978: organizes youth conferences on environmental issues such as biogas and alternatives to shifting cultivation.
50 **Indian Law Institute**, Delhi, 1956: publishes articles on pollution control, water resources management and law and politics.
51 **Indira Gandhi Institute for Development Research**, Mumbai, 1986: conducts multidisciplinary research and publishes papers on national resource accounting.
52 **Institute for Cultural Unity**, Kodaikanal (Tamil Nadu), 1987: promotes social forestry, waste land development and sustainable agricultural practices among small farmers.
53 **Institute for Integrated Rural Development**, Aurangabad (Maharashtra), 1987: organizes workshops on ecologically sound agriculture and women's environmental concerns (e.g. resisting chemical agriculture).
54 **Institution of Public Health Engineers in India**, Calcutta, 1973: promotes public awareness of issues relating to water supply, sanitation and afforestation.
55 **International Centre for Study and Development**, Valakom (Kerala), 1982: organizes environmental awareness campaigns on various issues such as water resources management.
56 **International Commission on Peace and Food**, Chennai, 1989: studies the connections between political, economic and ecological security in order to evolve strategies for accelerated sustainable development and jobs for the poor.
57 **International Institute of Sustainable Development and Management**, Ahmedabad (Gujarat), 1991: coordinates environmental research, training and the transfer of technology.
58 **International Society of Naturalists**, Baroda (Maharashtra), 1975: promotes wildlife conservation through education.
59 **Jagruthi**, Jeedimetla (Andhra Pradesh), 1989: studies water pollution in industrial areas and provides resource materials for education.

60 **Jana Seva Parishad** (People's Service Organization), Cuttack (Orissa), 1975: organizes self-help environmental schemes for afforestation, wildlife conservation and the use of renewable energy.
61 **Kerala Shastra Sahithya Parishad**, Trissur (Kerala), 1962: advocates environmentally sound sustainable development through biodiversity conservation, and the protection of freshwater and coastal areas.
62 **Kwality-Karnataka Welfare Society**, Chilbalapur (Karnataka), 1984: trains people through rural agricultural programmes.
63 **Ladakh Ecological Development Group**, Leh (Ladakh-Kashmir), 1983: promotes ecological and sustainable development which harmonizes with and builds on traditional culture.
64 **Lalbhai Group Rural Development Fund**, Ahmedabad (Gujarat), 1979: conducts training programmes on rural technology, social forestry, solar water pumps and cookers, and biomass projects.
65 **Maharishi Vedvyas Foundation for Studies in Cooperation**, Varanasi, 1978: conducts research and training on development cooperatives among the rural poor.
66 **National Foundation of Indian Engineers**, Delhi, 1987: carries out research into environmentally sound projects in transport, petrochemicals and electronics.
67 **Orissa Environmental Society**, Bhubaneswar (Orissa), 1982: promotes public awareness concerning the protection and conservation of environmental and natural resources.
68 **Orissa State Volunteers and Social Workers Association**, Puri (Orissa), 1980: promotes activities on soil conservation, afforestation, watershed development, rural energy and sustainable agriculture.
69 **Peace Trust – People's Education for Action and Community Emancipation**, Dindigul (Kerala), 1984: promotes local pollution prevention and public awareness raising on global warming and ozone depletion.
70 **People's Commission on Environment and Development**, Delhi, 1990: organizes public hearings for NGOs and independent individuals to express views on environmental issues and socio-economic development.
71 **Pradip Smriti Sansthan** (Pradip Memorial Organization), Patna (Bihar), 1982: promotes rural development and environmental management in Bihar.
72 **Prakruti** (Nature), Sirsi (Karnataka), 1983: encourages people's groups such as the Chipla Movement to involve rural populations in forest conservation.
73 **Prayog** (Usefulness), Kolar (Karnataka), 1986: educates to prevent soil erosion and encourage tree planting and the de-silting of water tanks.
74 **Rural Centre for Human Interests**, Rajgarh (Himachal Pradesh), 1980: promotes afforestation in mountain regions and the use of solar water heaters and cookers.

75 **Rural Development and Youth Training Institute**, Kota (Rajasthan), 1984: organizes training for rural youth and women in community development, healthcare and environmental protection (e.g. afforestation, the control of desertification).
76 **Rural Development Organization of India**, Manipur, 1975: organizes environmentally sensitive community development programmes for tribal people in the northeast.
77 **Ryan Foundation**, Chennai, 1990: conducts research into the desalination of sea water and brackish open and borehole water.
78 **Sadvichar Parivar**, Ahmedabad (Gujarat), 1965: conducts relief work in the aftermath of natural disasters (e.g. cyclones) and forestry programmes in collaboration with the State Forest Department.
79 **St Xavier's Social Service Society**, Ahmedabad (Gujarat), 1976: promotes appropriate technology (e.g. solar cookers) and social forestry through educational programmes.
80 **Sanskrit Shodh Sansthan** (Association for Studying Sanskrit), Rae Bareli (Uttar Pradesh), 1982: organizes meetings to encourage environmental awareness and traditional methods of communication.
81 **Sarita Society**, Udaipur (Rajasthan), 1991: deals with the problems of tribal people in Rajasthan's mineral deposits area (e.g. the effects of mining and ways in which traditional resources can improve their economic well-being).
82 **Save Nilgiris Campaign**, Chennai, 1986: mobilizes public opinion to protect the ecology of the Nilgiri Hills.
83 **School of Fundamental Research**, Calcutta, 1971: organizes environmental programmes for youth and others in collaboration with the Ministry of Environment and Forests.
84 **Sharmila Gramin Shilp Kala Kendara**, Patna (Bihar), 1985: organizes environmental education at village level in sanitation, biogas, tree planting and ecosystem preservation.
85 **Society for Human Integrity and Prosperity**, Nuzuid (Andhra Pradesh), 1986: collaborates with state government in repairing village water tanks, preventing the silting of rivers and similar problems.
86 **Society for Participatory Research in Asia**, Delhi, 1982: supports local initiatives in relation to water availability, social displacement due to large development projects and access to forest resources.
87 **Society for Rural Development**, Jaipur (Rajasthan), 1984: promotes environmental awareness through the training of social animators (primarily women and young people) and the cultivation of useful plant species.
88 **Sri Aurobindo Society**, Pondicherry, 1960: engages in research in natural technology (e.g. energy sources) and Indian culture.
89 **Sri Bajrang Lokmanas Kalyan Samiti**, Amargarh (Uttar Pradesh), 1986: promotes environmental education in sanitation, tree planting etc. among Adivasi tribal people.

90 **Taralabalu Rural Development Foundation**, Sirigere (Karnataka), 1982: promotes water conservation and afforestation and played a part in the development of the Earth Charter in Rio de Janeiro.
91 **Tata Energy Research Institute**, Delhi, 1974: promotes wide-ranging studies of energy options and their environmental consequences, including estimates of greenhouse gas emissions and policy responses for their minimization.
92 **Tropical Botanic Garden and Research Institute**, Thiruvananthapuram (Kerala), 1979: maintains a seed bank, herbarium and plant research institute designed to protect endangered tropical plant species and promote a variety of traditional plant usages and remedies.
93 **Trust for Transitional Action and Progress**, Tirupattur (Tamil Nadu), 1989: provides relief for the victims of natural disasters and ecological and health education.
94 **Trust-help**, Chennai, 1989: supports informal adult education centres catering for environmental, community health and literacy programmes.
95 **United Socio-economic Development and Research Programme**, Pune (Maharashtra), 1980: promotes slum improvement, rural development and biogas with special emphasis on the training of women and young people.
96 **Village Institution for Social Action**, Rourkela (Bihar), 1987: promotes self-reliant women's groups through the development of microenterprises and the conservation of local natural resources.
97 **Water Development Society**, Hyderabad, 1970: identifies groundwater potential areas, installs pumps and promotes environmental awareness in sanitation, water harvesting and soil erosion.
98 **Working Women's Forum**, Chennai, 1982: mobilizes women for collective action on family welfare, the problems of inter-caste marriages and child labour rehabilitation.
99 **Youth Charitable Organization**, Vizag (Andhra Pradesh), 1981: organizes self-help schemes for women and children in community health, forestry, animal husbandry, water resources and functional literacy.
100 **Yuva** (Youth), Ahmedabad (Gujarat), 1985: organizes voluntary work among young people in rural development and environmental protection, including water resources awareness, herbal gardens, alternative domestic fuels, tree planting and organic farming.

Select glossary

Throughout the book Hindu technical words are in Sanskrit. Buddhist ones may be in Sanskrit or Pali. In the case of word pairs, the Sanskrit comes first. Words which have been Anglicized (e.g. brahmin), are well known (e.g. karma), and the titles of sacred books are not italicized. Hindi words are rendered phonetically without diacritical marks. Terms used occasionally are explained in the text.

advaita	non-dual
ahiṁsā	non-injury
anātman, anattā	no-self
anicca	impermanent
āśrama	stage of life
ātman	self, spirit
avatāra	descent (e.g. as an embodied deity)
bhakti	devotion
bhikkhu	monk
bhikkhunī	ordained woman
bodhisattva	enlightenment being
Brahman	the supreme cosmic entity; literally 'the Great One'
brahmin	the highest of the four major categories of Vedic society
cakravartin	universal ruler
darśana	perspective, viewpoint
deva	the personification of transcendence
dev-van	sacred grove
dharma, dhamma	order, code of practice
duḥkha, dukkha	suffering, unsatisfactoriness
jāti	birth, socio-religious status
jīva	the living self
jñāna	knowing
kāma	desire, lust
karma	action
khandha	see *skandha*
kṣatriya	warrior, the second of the four major Vedic social groups

mae chii (Thai)	lay nun
masjid (Arabic)	mosque
māyā	supernatural power, illusoriness
mokṣa	liberation
nirvāṇa, nibbāna	the extinction of attachments
padayātrā	foot march
panchayat	council of five members
paṭicca samuppāda	interdependent co-arising
prajña	wisdom
prakṛti	primordial material nature
samādhi	the integrated state of mind which follows meditation
saṁsāra	cycle of existence
saṅgha	monastic community
sarvodaya	the awakening and welfare of all
satyāgraha	truth force
skandha, khandha	the aggregates which constitute human appearance
smṛti	remembered tradition
śruti	canonical tradition
stūpa	a monument based on religious relics
śūdra	menial, the fourth of the four major Vedic social groups
śūnyatā	emptiness, the absence of a permanent self
sūtra, sutta	authoritative text of aphorisms
tantra	esoteric ritual path
trikāya	the belief that the Buddha manifested himself in three bodies
vaiśya	trader, the third of the four major Vedic social groups
varṇa	colour, caste position
viññāna	consciousness, the highest of the five *khandhas*
vipassanā	insight meditation
wat (Thai)	temple or monastery
yajña	sacrificial rite
yoga	integrative discipline

Notes

1 Introduction

1. Sharon Beder, *Global Spin*, Totnes, UK, Green Books, 1997, and Vermont, USA, Chelsea Green Publishing Company, 1998, p. 161.
2. Jean Drèze and Amartya Sen, *India: Economic Development and Social Opportunity*, Delhi, Oxford University Press, 1996, p. 27.
3. Ibid., p. 78.
4. This figure was calculated by C. W. Hope of the Judge Institute of Management Studies, University of Cambridge.
5. *Our Common Future*, Report of the UN World Commission on Environment and Development, Oxford, Oxford University Press, 1987, p. 105.
6. Ibid., p. 111.
7. O. P. Dwivedi, *India's Environmental Policies, Programmes and Stewardship*, London, Macmillan, 1997.
    M. G. Rajan, *Global Environmental Politics*, Delhi, Oxford University Press, 1997.
    George A. James (ed.) *Ethical Perspectives on Environmental Issues in India*, Delhi, APH Publishing Corporation, 1999.
8. Indira Gandhi, 'Man and environment' (Plenary Session of UNCHE, 14 June 1972), in *Indira Gandhi on Environment*, New Delhi, Department of the Environment, Government of India, 1984, p. 20.
9. Vikram K. Akula, 'Grassroots environmental resistance in India', in Bron R. Taylor (ed.) *Ecological Resistance Movements*, Albany, State University of New York Press, 1995.
10. J. L. Chapman and M. J. Reiss, *Ecology*, Cambridge, Cambridge University Press, 1992, p. 3.
11. Chapman and Reiss, op. cit. (note 10).
    P. D. Sharma, *Ecology and Environment*, Meerut, Rastogi Publications, 1997.
    P. H. Collin, *Dictionary of Ecology and Environment*, Delhi, Universal Book Stall, 1996.
12. D. J. R. Angell, J. D. Comer and M. L. N. Wilkinson (eds) *Sustaining Earth*, London, Macmillan, 1990.
13. J. R. Engel and J. G. Engel (eds) *Ethics of Environment and Development*, London, Belhaven Press, 1990.
    L. S. Hamilton (ed.) *Ethics, Religion and Biodiversity*, Cambridge, White Horse Press, 1993.
14. W. Sachs (ed.) *Global Ecology: A New Arena of Political Conflict*, London, Zen Books, 1993.
15. N. Low (ed.) *Global Ethics and Environment*, London, Routledge, 1999.
    J. Connelly and G. Smith, *Politics and the Environment*, London, Routledge, 1999.

16 Christopher K. Chapple, 'Towards an indigenous Indian environmentalism', in Lance E. Nelson (ed.) *Purifying the Earthly Body of God*, Albany, State University of New York Press, 1998, p. 20.
17 R. Ninian Smart, in John Bowker (ed.) *The Oxford Dictionary of World Religions*, Oxford, Oxford University Press, 1997, p. xxiv.
18 Ṛgveda 5.53.9.
19 M. M. Thomas, *A Diaconal Approach to Indian Ecclesiology*, Rome, Centre for Indian and Inter-religious Studies, and Tiruvalla (Kerala), Christava Sahitya Samithy, 1995, p. 57.
20 Gandhi, op. cit. (note 8), p. 20.
21 J. B. Callicott, *Earth's Insights: A Survey of Ecological Ethics from the Mediterranean Basin to the Australian Outback*, Berkeley, University of California Press, 1994.
22 D. Kinsley, *Ecology and Religion: Ecological Spirituality in Cross-Cultural Perspective*, New Jersey, Prentice-Hall, 1995.
23 Chapple, op. cit. (note 16), p. 34.
24 Mary E. Tucker and Duncan R. Williams (eds) *Buddhism and Ecology*, Harvard, Harvard Center for the Study of World Religions, 1997.
25 Roger S. Gottlieb (ed.) *This Sacred Earth*, New York and London, Routledge, 1996, p. 10.
26 M. Marriott, 'Little communities in an indigenous civilization', in *Village India, Studies in the Little Community*, Chicago, Chicago University Press, 1955.
L. Dumont and D. Pocock (eds) *Contributions to Indian Sociology*; see especially 1957, no. 1; 1959, no. 3; and 1960, no. 4 .
27 Stanley J. Tambiah, *Buddhism and the Spirit Cults in North-east Thailand*, Cambridge, Cambridge University Press, 1970.
28 Jane Bunnag, *Journal of the Siam Society*, 1971, vol. 59, part 2, p. 278.
29 Tambiah, op. cit. (note 27), p. 370.
30 Ibid., p. 374.
31 David L. Gosling, *A New Earth*, London, Council of Churches for Britain and Ireland, 1992, p. 62.
32 Madhu Jain, 'God hasn't died young', in *India Today International*, 5 October 1998, p. 37.
33 Bridget and Raymond Allchin, *The Rise of Civilization in India and Pakistan*, Cambridge, Cambridge University Press, 1982, p. 3.
34 O. P. Dwivedi, *Environmental Crisis and Hindu Religion*, Delhi, Gitanjali Publishing House, 1987.
Ranchor Prime, *Hinduism and Ecology*, Motilal Bansaridass, Delhi, 1994.
35 David L. Gosling, *Science and Religion in India*, Series on Religion, no. 21, Bangalore, Christian Institute for the Study of Religion and Society; Madras, Christian Literature Society, 1976.
36 Madhav Gadgil and Ramachandra Guha, *Ecology and Equity*, London, Routledge, 1995; Delhi, Penguin Books, 1995, p. 123.
37 Drèze and Sen, op. cit. (note 2), p. 190.
38 W. Fernandez, 'Tribals, forests, displacement and sustainable development', in K. Gopal Iyer (ed.) *Sustainable Development: Ecological and Sociocultural Dimensions*, Delhi, Vikas Publishing House, 1996, p. 245.
39 David L. Gosling, 'Religious perspectives from Rio', in P. Doble and M. Hayward (eds) *The Contribution of Religious Education to Teaching about the Environment*, York, UK, York Religious Education Centre, University College of Ripon and York St John, 1993, p. 66.
40 Ibid., p. 67.
41 *Traditions, Concerns and Efforts in India, National Report to UNCED*, Ministry of Environment and Forests, Government of India, June 1992.

2 Ecology and Hindu tradition

1 Mahābhārata 1. 214–25.
2 Bridget and Raymond Allchin, *The Rise of Civilization in India and Pakistan*, Cambridge, Cambridge University Press, 1982, p. 3.
3 Vandana Shiva et al., *Ecology and the Politics of Survival*, Tokyo, United Nations University Press, and Delhi, Sage Publications, 1991, p. 75.
4 Madhav Gadgil and Ramachandra Guha, *This Fissured Land*, Delhi, Oxford University Press, 1992, p. 37.
5 Matthew Areeparampil, 'The religious aspect of the natural resource conflicts in the tribal society of Jharkand', paper submitted to ecology symposium at the Indian Social Institute, Delhi, 1996.
6 Allchin, op. cit. (note 2), p. 14.
7 Damodar D. Kosambi, *An Introduction to the Study of Indian History*, Bombay, Popular Prakashan, 1975, p. 66.
8 Gavin Flood, *Introduction to Hinduism*, Cambridge, Cambridge University Press, 1996.
9 Allchin, op. cit. (note 2), p. 225.
10 Śatapatha Brāhmaṇa 1.4.1.14–17.
11 Atharvaveda 12.1.3.
12 Ṛgveda 1.149.3.
13 Kosambi, op. cit. (note 7), p. 75. See especially Ṛgveda 2.15.8.
14 Ṛgveda 1.1.1.
15 Ṛgveda 10.7.3.
16 Ṛgveda 1.23.20.
17 Ṛgveda 2.21.6.
18 Ṛgveda 1.164.46.
19 Ṛgveda 10.90.12.
20 Yajurveda 13.27–9.
21 Yajurveda 37.17.
22 Yajurveda 32.8.
23 Yajurveda 40.6.
24 Bṛhad-āraṇyaka Upanishad 1.1.1.
25 Chāndogya Upanishad 6.12.3.
26 Bṛhad-āraṇyaka Upanishad 3.9.28.
27 Chāndogya Upanishad 6.11.1–2.
28 Śvetāśvatara Upanishad 6.7.
29 Julius Lipner, *Hindus: Their Religious Beliefs and Practices*, London, Routledge, 1994, p. 40.
30 Ibid., p. 176.
31 David L. Gosling, *Science and Religion in India*, Series on Religion, no. 21, Bangalore, Christian Institute for the Study of Religion and Society, and Madras, Christian Literature Society, 1976.
32 Lipner, op. cit. (note 29), p. 247.
33 Ṛgveda 1.65.7.
34 Kauṭilya, the Arthaśāstra, Adhyaksapracar 2.20.
35 Kauṭilya, the Arthaśāstra, 2.1.17.
36 Lipner, op. cit. (note 29), p. 162.
37 Gadgil and Guha, op. cit. (note 4), p. 103.
38 John Bowker (ed.), *The Oxford Dictionary of World Religions*, Oxford, Oxford University Press, 1997, p. 730.
39 Muṇḍaka Upanishad, 2.2.12.
40 Karan Singh, *Essays on Hinduism*, Delhi, Ratna Sagar, 1995, p. 140.
41 Chāndogya Upanishad 6.12.3.

42 Singh, op. cit. (note 40), p. 6.
43 Ibid.
44 Ibid., p. 141.
45 Bhagavadgītā 4.6–8.

3 Ecology and modern India

1 Madhav Gadgil and Ramachandra Guha, *This Fissured Land*, Delhi, Oxford University Press, 1993, p. 116.
2 Walter Fernandez, 'Tribals, forests, displacement and sustainable development', in K. Gopal Iyer (ed.) *Sustainable Development: Ecological and Sociocultural Dimensions,* Delhi, Vikas Publishing House, 1996, p. 259.
3 D. Brandis, *Indian Forestry*, Woking, UK, Oriental Institute, 1897, pp. 14–15.
4 David L. Gosling, 'Christian Response Within Hinduism', *Religious Studies*, 1974, vol. 10, no. 4, p. 433.
5 M. N. Srinivas, *Social Change in Modern India*, Berkeley, California University Press, 1968, p. 119.
6 David L. Gosling, *Science and Religion in India*, Bangalore, Christian Institute for the Study of Religion and Society, and Madras, Christian Literature Society, 1976, p. 7.
7 S. Mukherjee, *Hindustan Review*, May 1907, p. 480.
8 This encounter is apocryphal. It is echoed in several places in Vivekananda, *Complete Works,* vols I–X, Almora, Advaita Ashram, 1964.
9 Vivekananda, *Complete Works*, vol. V, Almora, Advaita Ashram, 1964, p. 519.
10 Gosling, op. cit. (note 6), p. 26.
11 Vivekananda, *Karma Yoga*, Almora, Advaita Ashram, 1930, pp. 131–2.
12 Basant Kumar Lal, *Contemporary Indian Philosophy,* Delhi, Motilal Bansaridass, 1995, p. 35.
13 Vivekananda, *Complete Works*, vol. II, Almora, Advaita Ashram, 1964, p. 414.
14 Ibid., vol. VI, p. 235.
15 M. M. Thomas, *The Acknowledged Christ of the Indian Renaissance*, Madras, Christian Literature Society, 1970.
16 Alec R. Vidler, *The Church in an Age of Revolution*, London, Penguin, 1961, p. 118.
17 Ibid., p. 117.
18 Gosling, op. cit. (note 6), p. 14.
19 P. C. Roy, *Life and Experiences of a Bengali Chemist*, London, Kegan Paul, Trench, Trübner & Co., Ltd, undated, p. 141.
The *Proceedings of the Decennial Missionary Conferences* from 1872 onwards were published in the year that they occurred.
20 J. C. Bose, 'Literature and science', in *Sir J. C. Bose, Life and Speeches*, Madras, Ganesh & Co., undated, p. 80.
21 P. C. Roy, *History of Hindu Chemistry*, vols I and II, Calcutta, Bengal Chemical and Pharmaceutical Works, 1902.
B. N. Seal, *The Positive Sciences of the Ancient Hindus,* London, Longman, Green, et al., 1915.
22 Roy, op. cit. (note 19), p. 42.
23 J. C. Bose, *Modern Review*, February 1917, vol. XXI, p. 202.
24 J. C. Bose, in M. Gupta, *Jagadish Chandra Bose: A Biography,* Bombay, Bharatiya Vidya Bhavan, 1964, p. 134.
25 Bose, op. cit. (note 20), p. 203.
26 J. C. Bose, 'Voice of life', *Modern Review*, vol. XXII, p. 590.
27 Bose, op. cit. (note 20), p. 42.
28 M. K. Gandhi, in V. B. Kher (ed.) *In Search of the Supreme*, vol. II, Ahmedabad, Navajivan Publishing House, 1961, p. 6.

29 Ibid., p. 5.
30 Ibid., p. 7.
31 M. K. Gandhi, *Young India*, 4 December 1924.
32 M. K. Gandhi, *An Autobiography*, Ahmedabad, Navajivan Publishing House, 1927, p. 198.
33 Ibid., p. 224.
34 M. K. Gandhi, *Speeches*, p. 276, quoted in Lal, op. cit. (note 12), p. 149.
35 Shriman Narayan (ed.) *The Selected Works of Mahatma Gandhi*, vol. VI, Ahmedabad, Navajivan Publishing House, 1968, p. 124.
36 Bhikhu Parekh, *Gandhi's Political Philosophy*, Delhi, Ajanta, 1989. See also Judith M. Brown, *Gandhi's Rise to Power*, Cambridge, Cambridge University Press, 1972.
37 Parekh, op. cit. (note 36), p. 196.
38 A. Coghlan, 'Sensitive flower', *New Scientist*, 26 September 1998, pp. 24–8.
39 Parekh, op. cit. (note 36), p. 109.

4 Struggles for the forests

1 Feisal Alkazi et al., *Chipko*, Delhi, Centre for Science and Environment, 1993, p. 18.
2 Sunderlal Bahuguna, *People's Programme for Change*, Environmental Series no. 19, Delhi, Indian National Trust for Art and Cultural Heritage, 1992, p. 7.
3 Ganesh and Vasudha Pangare, *From Poverty to Plenty: The Story of Ralegan Siddhi*, Studies in Ecology and Sustainable Development, no. 5, Delhi, Indian National Trust for Art and Cultural Heritage, 1992, p. xv.
4 Sunderlal Bahuguna, *Towards Basic Changes in Land Use*, Environmental Series no. 10, Delhi, Indian National Trust for Art and Cultural Heritage, 1989, p. 1.
5 S. M. Edwardes, 'Tree worship in India', *Empire Forestry*, 1922, vol. 1, no. 1, pp. 78–80.
6 Robert Redfield, *Peasant Society and Culture*, Chicago, Chicago University Press, 1961, pp. 19, 63–4.
7 Madhav Gadgil and Ramachandra Guha, *This Fissured Land*, Delhi, Oxford University Press, 1993, pp. 123–34.
8 Bahuguna, op. cit. (note 4), pp. 2–3.
9 Gadgil and Guha, op. cit. (note 7), p. 208–9.
10 Ramachandra Guha, *The Unquiet Woods*, Delhi, Oxford University Press, 1991, p. 93.
11 Ibid., p. 111.
12 Guha, op. cit. (note 10), p. 176.
13 Bahuguna, op. cit. (note 4), p. 6.
14 Ibid., p. 10.
15 Bahuguna, op. cit. (note 2), p. 5.
16 Ibid., p. 8.
17 Bahuguna, op. cit. (note 4), p. 11.
18 Bahuguna, op. cit. (note 2), p. 9.
19 Ibid., p. 7.
20 E. A. V. Prasad, *Ground Water in Varāhamihira's Bṛhat Saṁhitā*, Monographs in Ancient Scientific Sanskrit Literature, no. 1, Tirupati, Sri Venkateswara University Press, 1980.
21 Pangare, op. cit. (note 3), p. 6.
22 Ibid., p. 10.
23 T. N. Raghunatha, *The Pioneer*, 26 November 1996, p. 9.
24 Vandana Shiva et al., *Ecology and the Politics of Survival*, Tokyo, United Nations University Press, and Delhi, Sage Publications, 1991, p. 116.
25 David Kinsley, 'Learning the story of the land: reflections on the liberating power of geography and pilgrimage in the Hindu tradition', in Lance E. Nelson (ed.)

Purifying the Earthly Body of God, Albany, State University of New York Press, Albany, 1998, pp. 225–47.

5 Ecology and Buddhism

1. This interview was conducted with Dasho Paljor J. Dorji, deputy minister of the National Environment Commission of Bhutan, in his office in Thimpu on 3 October 1996.
2. These comments are based on a conversation with H.E. Om Pradhan at his home in Thimpu in October 1996.
3. Madhav Gadgil and Ramachandra Guha, *This Fissured Land*, Delhi, Oxford University Press, 1993, p. 83.
4. V. Fausboll (trans.), *Sutta-Nipāta*, Delhi, Motilal Bansaridass, 1968.
5. The Venerable Nanamoli Thera (trans.), *The Pāṭimokkha*, Bangkok, Mahamakut Academy, Science Association Press of Thailand, 1966, section 18.
6. Christopher K. Chapple, 'Animals and environment in the Buddhist birth stories', in Mary E. Tucker and Duncan R. Williams (eds) *Buddhism and Ecology*, Harvard, Harvard Center for the Study of World Religions, 1997, p. 145.
7. Aṅguttara Nikāya (Gradual Sayings), vol. 3, p. 262.
8. Gadgil and Guha, op. cit. (note 3), p. 88.
9. Damodar D. Kosambi, *An Introduction to the Study of Indian History*, Bombay, Popular Prakashan, 1975, p. 207.
10. F. L. Woodward, *The Book of Gradual Sayings*, vol. 2, 1933, p. 85; quoted in Trevor Ling, *The Buddha*, London, Temple Smith, 1973, p. 142.
11. John Bowker (ed.) *The Oxford Dictionary of World Religions*, Oxford, Oxford University Press, 1997, p. 930.
12. Siksasamuccaya, p. 278, in W. T. de Bary (ed.) *Sources of Indian Tradition*, vol. I, New York, Columbia University Press, 1958, p. 161.
13. Trevor Ling, *The Buddha*, London, Temple Smith, 1973, p. 175.
14. Ibid., p. 193.
15. Daisaku Ikeda, *The Flower of Chinese Buddhism*, New York and Tokyo, Weatherhill, 1986, p. 33.
16. Helena Norberg-Hodge, *Ancient Futures*, Delhi, Oxford University Press, 1992, p. 73.
17. Ibid., p. ix.
18. Helena Norberg-Hodge, 'May a hundred plants grow from one seed', in M. Batchelor and K. Brown (eds) *Buddhism and Ecology*, Delhi, Motilal Bansaridass, 1994, p. 54.
19. This discussion took place at the Government College, Leh, in October 1995.
20. P. P. Karan, *Bhutan: Environment, Culture and Development Strategy*, Delhi, Intellectual Publishing House, 1990, p. 87.
21. David L. Gosling, 'The scientific and religious beliefs of Thai scientists and their inter-relationship', *Southeast Asian Journal of Social Science*, 1975, vol. 4, no. 1, pp. 1–18.
22. Donald Swearer, 'The hermeneutics of Buddhist ecology', in Tucker and Williams, op. cit. (note 6), p. 28.

6 Thailand: a case study

1. A Thai monk should be greeted by folding the hands under the chin and lowering one's eyes. The hand gesture is known as a *wai*.
2. Michael J. G. Parnwell and Raymond L. Bryant (eds) *Environmental Change in South-East Asia*, London, Routledge, 1996, p. 6. More generally, see also Philip Hirsch and Carol Warren (eds) *The Politics of Environment in Southeast Asia*, London, Routledge, 1998.

3 Jonathan Rigg (ed.) *Counting the Costs: Economic Growth and Environmental Change in Thailand*, Singapore, Institute of Southeast Asian Studies, 1995, p. 13.
4 Stanley J. Tambiah, *World Conqueror and World Renouncer*, Cambridge, Cambridge University Press, 1976, p. 185.
5 Ibid., p. 102.
6 Ibid., p. 401.
7 David L. Gosling, 'Thai Buddhism in transition', *Religion*, 1977, vol. 7, no. 1, pp. 18–34.
8 Santikaro Bhikkhu, 'Buddhadāsa Bhikkhu', in Christopher S. Queen and Sallie B. King (eds) *Engaged Buddhism*, Albany, State University of New York Press, 1996, p. 160.
9 Donald K. Swearer, 'The hermeneutics of Buddhist ecology', in Mary E. Tucker and Duncan R. Williams (eds) *Buddhism and Ecology*, Harvard, Harvard Center for the Study of World Religions, 1997, pp. 26–7.
10 Ibid., p. 34.
11 Tambiah, op. cit. (note 4), p. 429.
12 Ampa Santimetaneedol, 'Facing death with dignity', *Bangkok Post*, 6 November 1991.
13 Somboon Suksamran, *Political Buddhism in Southeast Asia*, London, St Martin's Press, 1976, p. 104. See also, same author, *Buddhism and Politics in Thailand*, Singapore, Institute of Southeast Asian Studies, 1982.
14 David L. Gosling, 'New directions in Thai Buddhism', *Modern Asian Studies*, 1980, vol. 14, no. 3, p. 415.
15 David L. Gosling, 'Thai monks in rural development', *Southeast Asian Journal of Social Science*, 1981, vol. 9, nos. 1–2, pp. 78–85.
16 William J. Klausner, *Reflections on Thai Culture*, Bangkok, Siam Society, 1993.
17 David L. Gosling, 'Redefining the saṅgha's role in Northern Thailand: an investigation of monastic careers at five Chiang Mai wats', *Journal of the Siam Society*, 1983, vol. 71, parts 1 and 2, p. 94.
18 David L. Gosling, 'Biogas for rural development: transferring the technology', *Biomass*, 1982, vol. 2, no. 4, pp. 309–16. See also, same author, *Vitritakan Palangngan nai Prathet Thai le Asiatawanogchiengtai* (Energy Crisis in Thailand and Southeast Asia), Bangkok, Komol Keemthong Foundation, 1981.
19 Gosling, op. cit. (note 14), p. 431.
20 David L. Gosling, 'Visions of salvation: a Thai Buddhist experience of ecumenism', *Modern Asian Studies*, 1992, vol. 26, no. 1, p. 37.
21 David L. Gosling, 'Thailand's bare-headed doctors', *Modern Asian Studies*, 1985, vol. 19, no. 4, p. 793. See also, same author, *Maw Phra* (Doctor-Monk), Bangkok, Komol Keemthong Foundation, 1986.
22 David L. Gosling, 'Thailand's bare-headed doctors: Thai monks in rural health care', *Journal of the Siam Society*, 1986, vol. 74, pp. 83–106, and in shortened form in *Journal of the National Research Council of Thailand*, 1987, vol. 19, no. 1, part II, pp. 1–10.
23 David L. Gosling, 'Thai monks and lay nuns (*mae chii*) in urban health care', *Anthropology and Medicine*, 1998, vol. 5, no. 1, pp. 5–23.
24 Donald K. Swearer, 'Sulak Sivaraksa's vision for renewing society', in Queen and King, op. cit. (note 8), p. 210.
25 Ibid., p. 222.
26 'Sulak steps up pipeline campaign', *Bangkok Post*, 2 August 1998, p. 1.
27 Vasana Chinvarakorn, 'Dawn of a new age', *Bangkok Post*, 'Outlook', 6 December 1999, p. 1.
28 Ibid.
29 David L. Gosling, 'Urban Thai Buddhist attitudes to development', *Journal of the Siam Society*, 1996, vol. 84, part 2, pp. 103–20.

30 Charles F. Keyes, 'Political crisis and militant Buddhism in contemporary Thailand', in Bardwell L. Smith (ed.) *Religion and Legitimation of Power in Thailand, Laos and Burma*, Chambersberg, Pennsylvania, Anima, 1978, p. 149.
31 'Scholar demands action over Dhammachayo letter', *Bangkok Post*, 1 May 1999.
32 Peter A. Jackson, *Buddhism, Legitimation and Conflict*, Singapore, Institute of Southeast Asian Studies, 1989, p. 212.
33 Phichai Tovivich, 'Monosodium glutamate: poisoning food for profit', in David L. Gosling and Feliciano V. Cariño (eds) *Technology from the Underside*, Geneva, World Council of Churches and Quezon City, National Council of Churches of the Philippines, 1986, pp. 87–92.
34 Khin Thitsa, *Providence and Prostitution*, Change International Reports, London, Parnell House, 1980, p. 4.
35 Nanthana Chaiyasut (ed.) *Report on a Survey of the Status of Women in Two Provinces,* Bangkok, publisher unknown, 1977, p. 26.
36 Chatsumarn Kabilsingh, *Thai Women in Buddhism*, Berkeley, Parallax Press, 1991, p. 19.
37 Ibid.
38 A. Thomas Kirsch, 'Economy, polity and change in Thailand', in G. W. Skinner and Thomas Kirsch (eds) *Change and Persistence in Thai Society*, Ithaca, Cornell University Press, 1975, pp. 172–96.
39 Thitsa, op. cit. (note 34), p. 23.
40 Charles F. Keyes, 'Mother or mistress but never a monk: Buddhist notions of female gender in rural Thailand', *American Ethnologist*, 1984, vol. 11, part 2, p. 224.
41 Kabilsingh, op. cit. (note 36), p. 25.
42 Nancy J. Barnes, 'Buddhist women and nuns' order in Asia', in Queen and King, op. cit. (note 8), p. 261.
43 Susan Murcott, *The First Buddhist Women*, Berkeley, Parallax Press, 1991, p. 10.
44 Kabilsingh, op. cit. (note 36), p. 36.
45 Samer Boonma, 'Bhikkhunī in Buddhism', unpublished MA thesis, Bangkok, Chulalongkorn University, 1978, pp. 114–25.
46 *The Rules of the Foundation of the Nun Institute of Thailand*, Bangkok, Suntsiri Press, 1979, p. 35.
47 David L. Gosling, 'The changing roles of Thailand's lay nuns (*mae chii*)', *Southeast Asian Journal of Social Science*, 1998, vol. 26, no. 1, pp. 121–43; see also, same author, op. cit. (note 23), p. 133.
48 David L. Gosling, 'Buddhism for peace', *Southeast Asian Journal of Social Science*, 1984, vol. 12, no. 1, pp. 59–70.

7 India since Independence

1 Indian Constitution in the *Manorama Yearbook*, Kottayam, Malayala Manorama, 1998, p. 468.
2 Trevor Ling, *The Buddha*, London, Maurice Temple Smith, 1973, reprinted by Lowe and Brydone, 1974, p. 243.
3 *Manorama Yearbook*, op. cit. (note 1), p. 471.
4 Ayesha Jalal, *Democracy and Authoritarianism in South Asia*, Cambridge, Cambridge University Press, 1995, p. 245.
5 Jean Drèze and Amartya Sen, *India: Economic Development and Social Opportunity*, Delhi, Oxford University Press, 1996, p. 10.
6 Jawaharlal Nehru, Constituent Assembly, Delhi, 1947, reprinted in Sarvepalli Gopal (ed.) *Jawaharlal Nehru: An Anthology*, Delhi, Oxford University Press, 1983.
7 Cécile de Sweemer, 'The Bhopal industrial disaster', in David L. Gosling and Feliciano V. Cariño (eds) *Technology from the Underside*, Geneva, World Council

of Churches, and Quezon City, National Council of Churches in the Philippines, 1986, p. 58.
8   Madhav Gadgil and Ramachandra Guha, *Ecology and Equity*, London, Routledge, 1995, and Delhi, Penguin Books, 1995, p. 13.
9   Ibid.
10 Ibid., p. 36.
11 Ibid., p. 50.
12 *Traditions, Concerns and Efforts in India, National Report to UNCED*, Delhi, Ministry of Environment and Forests, Government of India, June 1992.
13 Indira Gandhi, 'Man and environment' (Plenary Session of UNCHE, 14 June 1972), in *Indira Gandhi on Environment*, Delhi, Department of the Environment, Government of India, 1984, pp. 20–9.
14 Ibid., p. 29.
15 Ibid., p. 27.
16 Sandria B. Freitag, 'Contesting in public', in David Ludden (ed.) *Making India Hindu*, Delhi, Oxford University Press, 1997, p. 218.
17 Tapan Basu et al., *Khaki Shorts and Saffron Flags*, Tracts for the Times, no. 1, Delhi, Orient Longman, 1993, p. 32.
18 Ibid., p. 48.
19 Ibid., p. 71.
20 Ibid., p. 33.
21 Ibid.
22 Gadgil and Guha, op. cit. (note 8).
23 Drèze and Sen, op. cit. (note 5).
24 Gadgil and Guha, op. cit. (note 8), p. 115.
25 Ibid., p. 123.
26 Ibid.
27 Drèze and Sen, op. cit. (note 5), p. 182.
28 Ibid., p. 79.
29 Ibid., p. 14.
30 Ibid., p. 39.
31 Ibid., p. 91.
32 Ibid., p. 185.
33 Ibid., p. 107.
34 Ibid., p. 190.

8 Signs of hope

1   Tavleen Singh, 'Beware the eco-terrorist', *India Today*, 17 November 1997, p. 57.
2   Medha Patkar, 'Dam and dissent', *India Today*, 27 July 1998, p. 27.
3   Tavleen Singh, 'Luddite sisters', *India Today*, 22 June 1998, p. 9.
4   *Statement to the United Nations Conference on Environment and Development*, Brahma Kumaris World Spiritual University, June 1992.
5   Sunderlal Bahuguna, 'The cry of the Himalayas', *Hindustan Times*, 5 June 1998, p. 6.
6   S. Singh, 'Environment, class and state in India: a perspective on sustainable irrigation', unpublished PhD dissertation, Department of Political Science, University of Delhi, 1994, p. 259.
7   'Is the Tehri Dam safe?', special report, *Down to Earth*, 15 June 1998, p. 20.
8   Sunderlal Bahuguna, 'Our Himalayan blunders', interview in *Sunday Times of India*, 1 March 1998.
9   Sudha Mahalingam, 'End of another round', *Frontline*, 26 December 1997, p. 98.
10 'Hindu *sammelan* opposes Tehri project', *Times of India*, 23 October 1997.
11 'Tactical withdrawal', *Indian Express*, 15 April 1998.

12 Raj Chengappa, 'Angels of change', *India Today*, 15 January 1996, p. 79.
13 Ibid., p. 80.
14 Uday Mahurkar, 'The fire inside', *India Today*, 15 February 1997, p. 123.
15 Uday Mahurkar, 'Saurashtra's rainmaker', *India Today*, 23 June 1997, p. 90. See also, same author, 'How green is my village', *India Today*, 11 September 2000, p. 58.
16 'Sound service', *India Today*, 20 October 1997, p. 22.
17 B. K. Sinha, 'The answers within', *Down to Earth*, 31 May 1998, p. 42.
18 Ranchor Prime, 'Saving Krishna's forests', *Down to Earth*, 31 March 1998, p. 38.
19 Darab J. Nagarwalla, 'The Himalayan herb grower', *Down to Earth*, 15 May 1999, p. 39.
20 'Waste venture', *Down to Earth*, 15 May 1997, p. 16.
21 Rasheeda Bhagat, 'Testing out a development project', *Frontline*, 26 December 1997, p. 47.
22 Amarnath K. Menon, 'Potter with a green dream', *India Today*, 4 August 1997, p. 84.
23 Saritha Rai, 'Medic with a mission', *India Today*, 7 July 1997, p. 78.
24 E. R. C. Davidar, *Cheetal Walk: Living in the Wilderness*, Delhi, Oxford University Press, reviewed by Tariq Aziz in *India Today*, 12 January 1998, p. 81.
25 Rohit Brijnath, 'Voice of the tiger', *India Today*, 31 December 1995, p. 196.
26 Madhav Gadgil, 'Making of a naturalist', *Frontline*, 6 March 1998, p. 68.
27 Avelin Mary, communication from the International Biographical Research Centre, Cambridge, Massachusetts, 24 August 1998.
28 'People power', *Down to Earth*, 15 June 1998, p. 38.
29 'Making power', *Down to Earth*, 15 June 1998, p. 50.
30 'Polluters are today's environmentalists', *Down to Earth*, 15 March 1998, p. 55.
31 M. G. Radhakrishnan, 'Friend of the fisherfolk', *India Today*, 10 November 1997, p. 70.
32 Interviews conducted on 13 March 1996 and 30 March 1999 at the India International Centre, Delhi.
33 Soli J. Sorabjee, 'Judicial activism', *Manorama Yearbook*, Kottayam, Malayala Manorama, 1998, p. 478.
34 Interview conducted on 15 May 1998 at the home of M. C. Mehta in Delhi.
35 Amartya K. Sen, 'Consequential evaluation and practical reason', April 2000, unpublished, p. 6.
36 Ibid., p. 11.
37 Ibid., p. 12.
38 Ibid., p. 30.
39 Roderick Hindery, *Comparative Ethics in Hindu and Buddhist Traditions*, Delhi, Motilal Bansaridass, 1978, 1996, pp. 128–56.
40 Anil Agarwal, 'The poverty of Amartya Sen', *Down to Earth*, 15 December 1998, p. 56.
41 Interview conducted on 10 May 2000 at Trinity College, Cambridge.
42 Interview conducted on 1 June 1998 at the home of Mrs Neelam Dewan, Shimla.
43 Interview with Vijay Paranjpye and visit took place on 18 March 1996 (Pune).

9 Expanding our horizons

1 Robert W. Bradnock, 'Glacial response to global warming claim', *The Guardian*, 4 February 2000.
2 Cécile de Sweemer gave her views to me in December 1999 in Vientiane, Laos.
3 Raja Rao, *The Serpent and the Rope*, Delhi, Hind Pocket Books and Orient Paperbacks, 1948, p. 207. The French translates as: 'God dwells in the space between men'.
4 This frequently used Thai phrase means 'never mind'.

5 Harold Coward, 'The ecological implications of karma theory', in Lance E. Nelson (ed.) *Purifying the Earthly Body of God*, Albany, State University of New York Press, 1998, pp. 39–51.
6 David L. Gosling, 'Science and Indian religion', *Religion and Society*, 1998, vol. 45, no. 2, p. 29.
7 Bridget and Raymond Allchin, *The Rise of Civilization in India and Pakistan*, Cambridge, Cambridge University Press, 1982, p. 3.
8 Julius Lipner, *Hindus: Their Religions, Beliefs and Practices*, London, Routledge, 1994, p. 177.
9 Lance E. Nelson, 'The dualism of nondualism: Advaita Vedānta and the irrelevance of nature', in Nelson, op. cit. (note 5), p. 61.
10 David L. Gosling, *Science and Religion in India*, Series on Religion, no. 21, Bangalore, Christian Institute for the Study of Religion and Society, and Madras, Christian Literature Society, 1976, pp. 71–5.
11 Walpola Rahula, *History of Buddhism in Ceylon*, Colombo, published by Walpola Rahula, second edition, 1966, p. 76.
12 Trevor Ling, *The Buddha*, London, Temple Smith, 1973, reprinted 1974, pp. 240, 239.
13 David Kinsley, 'Learning the story of the land: reflections on the liberating power of geography and pilgrimage in the Hindu tradition', in Nelson, op. cit. (note 5), p. 235.
14 Donald K. Swearer, 'The hermeneutics of Buddhist ecology', in Mary E. Tucker and Duncan R. Williams (eds) *Buddhism and Ecology*, Harvard, Harvard Center for the Study of World Religions, 1997, p. 21.
15 Vandana Shiva, *Ecology and the Politics of Survival*, Tokyo, United Nations University Press, and Delhi, Sage Publications, 1991, p. 50.
16 Tom Wheldon, 'The accountant's delusion', *Physics World*, June 1999, p. 72.
17 Gosling, op. cit. (note 10), p. 31.
18 *Our Common Future*, Report of the UN World Commission on Environment and Development, Oxford, Oxford University Press, 1987, p. 138.
19 Kenneth Kraft, 'Nuclear ecology and engaged Buddhism', in Tucker and Williams, op. cit. (note 14), p. 269.
20 Patrick Moore in 'Opinion interview', *New Scientist*, 25 December 1999, pp. 74–7.
21 Richard Jefferson in 'A helping hand?', *The Editor*, London, Guardian Newspapers, 31 March 2000, p. 13.
22 Martin Palmer et al. (eds) *Faith and Nature*, London, World Wide Fund for Nature and ICOREC, undated.

Select bibliography

Some titles are included which are not mentioned in the text.

Akula, Vikram K., 'Grassroots environmental resistance in India', in Bron R. Taylor (ed.) *Ecological Resistance Movements*, Albany, State University of New York Press, 1995.
Allchin, Bridget and Allchin, Raymond, *The Rise of Civilization in India and Pakistan*, Cambridge, Cambridge University Press, 1982.
Angell, D. J. R., Comer, J. D. and Wilkinson, M. L. N. (eds) *Sustaining Earth*, London, Macmillan, 1990.
Bahuguna, Sunderlal, *People's Programme for Change*, Environmental Series no. 19, Delhi, Indian National Trust for Art and Cultural Heritage, 1992.
Basu, Tapan et al., *Khaki Shorts and Saffron Flags*, Tracts for the Times, no. 1, Delhi, Orient Longman, 1993.
Beder, Sharon, *Global Spin*, Totnes, UK, Green Books, 1997, and Vermont, USA, Chelsea Green Publishing Company, 1998.
Birch, Charles, *Regaining Compassion*, Kensington, New South Wales University Press, 1992.
Bowker, John (ed.) *The Oxford Dictionary of World Religions*, Oxford, Oxford University Press, 1997.
Brown, J. M., *Gandhi: Prisoner of Hope*, New Haven and London, Yale University Press, 1989.
Callicott, J. B., *Earth's Insights: A Survey of Ecological Ethics from the Mediterranean Basin to the Australian Outback*, Berkeley, University of California Press, 1994.
Chapman, J. L. and Reiss, M. J., *Ecology*, Cambridge, Cambridge University Press, 1992.
Chapple, Christopher Key and Evelyn Tucker (eds) *Hinduism and Ecology: The Intersection of Earth, Sky and Water*, Harvard, Harvard University Press, 2000.
Collin, P. H., *Dictionary of Ecology and Environment*, Delhi, Universal Book Stall, 1996.
Connelly, J. and Smith, G., *Politics and the Environment*, London, Routledge, 1999.
Drèze, Jean and Sen, Amartya, *India: Economic Development and Social Opportunity*, Delhi, Oxford University Press, 1996.
Dwivedi, O. P., *Environmental Crisis and Hindu Religion*, Delhi, Gitanjali Publishing House, 1987.
—— *India's Environmental Policies, Programmes and Stewardship*, London, Macmillan, 1997.
Engel, J. R. and Engel, J. G. (eds) *Ethics of Environment and Development*, London, Belhaven Press, 1990.

Flood, Gavin, *Introduction to Hinduism*, Cambridge, Cambridge University Press, 1996.
Gadgil, Madhav and Guha, Ramachandra, *This Fissured Land*, Delhi, Oxford University Press, 1992.
—— *Ecology and Equity*, London, Routledge, 1995; Delhi, Penguin Books, 1995.
Ganeri, Jonardon, *Philosophy in Classical India: An Introduction and Analysis*, Routledge, 2001.
Ghosananda, Maha, *Step by Step*, Berkeley, Parallax Press, 1992.
Gosling, David L., 'The scientific and religious beliefs of Thai scientists and their inter-relationship', *Southeast Asian Journal of Social Science*, 1975, vol. 4, no. 1, pp. 1–18.
—— *Science and Religion in India*, Series on Religion, no. 21, Bangalore, Christian Institute for the Study of Religion and Society; Madras, Christian Literature Society, 1976.
—— 'Thai Buddhism in transition', *Religion*, 1977, vol. 7, no. 1, pp. 18–34.
—— 'New directions in Thai Buddhism', *Modern Asian Studies*, 1980, vol. 14, no. 3, pp. 411–39.
—— 'Thai monks in rural development', *Southeast Asian Journal of Social Science*, 1981, vol. 9, nos. 1–2, pp. 74–85.
—— *Vitritakan Palangngan nai Prathet Thai le Asiatawanogchiengtai* (Energy Crisis in Thailand and Southeast Asia), Komol Keemthong Foundation, 1981.
—— 'Biogas for rural development: transferring the technology', *Biomass*, 1982, vol. 2, no. 4, pp. 309–16.
—— 'Redefining the *saṅgha*'s role in northern Thailand: an investigation of monastic careers at five Chiang Mai wats', *Journal of the Siam Society*, 1983, vol. 71, parts 1 and 2, pp. 89–120.
—— 'Buddhism for peace', *Southeast Asian Journal of Social Science*, 1984, vol. 12, no. 1, pp. 59–70.
—— 'Thailand's bare-headed doctors', *Modern Asian Studies*, 1985, vol. 19, no. 4, pp. 761–96.
—— *Maw Phra* (Doctor Monk), Komol Keemthong Foundation, 1986.
—— 'Thailand's bare-headed doctors: Thai monks in rural health care', *Journal of the Siam Society*, 1986, vol. 74, pp. 83–106, and in shortened form in *Journal of the National Research Council of Thailand*, 1987, vol. 19, no. 1, part II, pp. 1–10.
—— *A New Earth*, London, Council of Churches for Britain and Ireland, 1992.
—— 'Visions of salvation: a Thai Buddhist experience of ecumenism', *Modern Asian Studies*, 1992, vol. 26, no. 1, pp. 31–47.
—— 'Religious perspectives from Rio', in P. Doble and M. Hayward (eds) *The Contribution of Religious Education to Teaching about the Environment*, York, UK, York Religious Education Centre, University College of Ripon and York St John, 1993, pp. 61–71.
—— 'Urban Thai Buddhist attitudes to development', *Journal of the Siam Society*, 1996, vol. 84, part 2, pp. 103–21.
—— 'Thai monks and lay nuns (*mae chii*) in urban health care', *Anthropology and Medicine*, 1998, vol. 5, no. 1, pp. 5–23.
—— 'The changing roles of Thailand's lay nuns (*mae chii*)', *Southeast Asian Journal of Social Science*, 1998, vol. 26, no. 1, pp. 121–43.
—— 'Science and Indian religion', *Religion and Society*, 1998, vol. 45, no. 2, pp. 5–31.
—— and Cariño, Feliciano V. (eds) *Technology from the Underside*, Geneva, World

Council of Churches and Quezon City, National Council of Churches of the Philippines, 1986.
Gottlieb, Roger S. (ed.) *This Sacred Earth: Religion, Nature, Environment*, New York and London, Routledge, 1996.
Gross, Rita, *Soaring and Settling: Buddhist Perspectives on Contemporary Social and Religious Issues*, London, Continuum, 1999.
Grove, Richard H., *Ecology, Climate and Empire: The Indian Legacy in Global Environmental History 1400–1940*, Delhi, Oxford University Press, 1998.
Guha, Ramachandra, *The Unquiet Woods*, Delhi, Oxford University Press, 1991.
Hamilton, L. S. (ed.) *Ethics, Religion and Biodiversity*, Cambridge, White Horse Press, 1993.
Hindery, Roderick, *Comparative Ethics in Hindu and Buddhist Traditions*, Delhi, Motilal Bansaridass, 1978, 1996.
Hirsch, Philip and Warren, Carol (eds) *The Politics of Environment in Southeast Asia*, London, Routledge, 1998.
Iyer, K. Gopal (ed.) *Sustainable Development: Ecological and Sociocultural Dimensions*, Delhi, Vikas Publishing House, 1996.
Jalal, Ayesha, *Democracy and Authoritarianism in South Asia*, Cambridge, Cambridge University Press, 1995.
James, George A. (ed.) *Ethical Perspectives on Environmental Issues in India*, Delhi, APH Publishing Corporation, 1999.
Kabilsingh, Chatsumarn, *Thai Women in Buddhism,* Berkeley, Parallax Press, 1991.
Keyes, Charles F., 'Mother or mistress but never a monk: Buddhist notions of female gender in rural Thailand', *American Ethnologist,* 1984, vol. 11, part 2.
Kinsley, D., *Ecology and Religion: Ecological Spirituality in Cross-Cultural Perspective*, New Jersey, Prentice-Hall, 1995.
Klausner, William J., *Reflections on Thai Culture*, Bangkok, Siam Society, 1993.
Lipner, Julius, *Hindus: Their Religious Beliefs and Practices*, London, Routledge, 1994.
Low, N. (ed.) *Global Ethics and Environment*, London, Routledge, 1999.
Ludden, David (ed.) *Making India Hindu*, Delhi, Oxford University Press, 1997.
Murcott, Susan, *The First Buddhist Women,* Berkeley, Parallax Press, 1991.
Nath, Vikas, *Sacred Groves: An Emerging Ethno-Religious Concept for Conservation of Plant Resources*, Delhi, Agricultural Finance Consultants, 1997.
Nelson, Lance E. (ed.) *Purifying the Earthly Body of God: Religion and Ecology in Hindu India*, Albany, State University of New York Press, 1998.
Norberg-Hodge, Helena, *Ancient Futures*, Delhi, Oxford University Press, 1992.
Palmer, Martin et al. (eds) *Faith and Nature*, London, World Wide Fund for Nature and ICOREC, undated.
Pangare, Ganesh and Vasudha*, From Poverty to Plenty: The Story of Ralegan Siddhi*, Studies in Ecology and Sustainable Development, no. 5, Delhi, Indian National Trust for Art and Cultural Heritage, 1992.
Parekh, Bhikhu, *Gandhi's Political Philosophy*, Delhi, Ajanta, 1989.
Parnwell, Michael J. G. and Bryant, Raymond L. (eds) *Environmental Change in South-East Asia*, London, Routledge, 1996.
Parpola, Asko and Hansen, Bent Smidt (eds) *South Asian Religion and Society*, London, Curzon Press, 1986.
Queen, Christopher S. and King, Sallie B. (eds) *Engaged Buddhism*, Albany, State University of New York Press, 1996.

Rajan, M. G., *Global Environmental Politics*, Delhi, Oxford University Press, 1997.
Rajotte, Freda, *First Nations Faith and Ecology*, Toronto, Cassell, 1998.
Rigg, Jonathan (ed.) *Counting the Costs: Economic Growth and Environmental Change in Thailand*, Singapore, Institute of Southeast Asian Studies, 1995.
Sachs, W. (ed.) *Global Ecology: A New Arena of Political Conflict*, London, Zen Books, 1993.
Sadahat, John, *Ways to Meaning and a Sense of Universality*, Mississauga, Ontario, Canadian Educators' Press, 1998.
Schader-Frenchette, R. S., *Environmental Ethics*, Boxwood, California, Pacific Grove, 1981.
Sen, Amartya K., 'Consequential evaluation and practical reason', unpublished essay, April 2000.
Shiva, Vandana et al., *Ecology and the Politics of Survival*, Tokyo, United Nations University Press, and Delhi, Sage Publications, 1991.
Singh, Karan, *Essays on Hinduism*, Delhi, Ratna Sagar, 1987, 1995.
Smith, Bardwell L. (ed.) *Religion and Legitimation of Power in Thailand, Laos and Burma*, Chambersberg, Pennsylvania, Anima, 1978.
Suksamran, Somboon, *Political Buddhism in Southeast Asia*, London, St Martin's Press, 1976.
Tambiah, Stanley J., *World Conqueror and World Renouncer*, Cambridge, Cambridge University Press, 1976.
Traditions, Concerns and Efforts in India, National Report to UNCED, Ministry of Environment and Forests, Government of India, June 1992.
Tucker, Mary E. and Williams, Duncan R. (eds) *Buddhism and Ecology: The Interconnection of Dharma and Deeds*, Harvard, Harvard Center for the Study of World Religions, 1997.
Williams, Paul, *Buddhist Thought*, London, Routledge, 2000.

Index

advaita 26, 38, 46, 166
Advani, L. K. 124; *see also* Hindu Right
Agni 10, 16, 27, 32, 37, 69, 163; in Śatapatha Brāhmaṇa 21–3
ahiṁsā 46, 47, 48, 50, 166
AIDS helpline 106
ākāśa 38, 42
Ali, Salim 147
Allchin, Bridget and Raymond 10, 17, 20, 32, 162
Ambedkar, B. R. 111
Amte, Baba 141
Angell, D. J. R. 5
Aṅguttara Nikāya 71
anicca 70
Antala, Shyami 144
antaryāmin 26
Apfell-Marglin, Frédérique 6
Appiko 4, 14, 51, 59, 60–1, 174
arhat 73
Ariyaratne, A. T. 166
Arjuna: in Bhagavadgītā 16, 153–4, 175; burning of Khāṇḍava forest 11, 32, 148, 149, 151, 163
Arthaśāstra 27, 28, 63, 71, 157
Aryans 21, 27, 69
Arya Samaj 37, 122, 123, 164
Ashoka: conservation policy 8, 14, 71–2; cosmic role 12, 84,135, 164, 165; legacy of 89–91; missionary outreach 74, 76
āśrama 27
Asur legend 19, 32
artharvan 23
Atharvaveda 22, 149
ātman 11, 24, 25, 26, 30, 32, 71, 163
atmosphere 5
Avalokiteśvara 73, 76
avatāra 33
Azariah, K. 146

Babri *masjid* 114, 124, 141; *see also* Islam
Badoni, Arun Kumar 145
Bahuguna, Sunderlal 53, 54, 61–3, 148, 166, 170, 171; national activities 13, 59, 66, 140, 141; pilgrimages 52, 167; work with Chipko movement 11, 51, 67, 156; *see also* Chipko; *padayātrā*
Bangladesh 137
Basu, Tapan 125, 126
Bedi, Kiran 145
Behn, Mira 61
Bhaba, Homi J. 43
Bhagavadgītā 16, 156, 163; consequential evaluation of 153–4, 175; influence on Gandhi 46, 49; *niṣkāma* karma (selfless action) 31, 33, 63, 151; use by Chipko leaders 11, 59, 66
Bharatiya Janata Party (BJP) 12, 38, 65, 124, 125
Bhatt, Chandi Prasad 61, 141; leader of Chipko 51, 59, 63; use of appropriate technology 52, 66, 67, 139, 171; *see also* Chipko
bhikkhunī 103, 105
Bhopal disaster 4, 127, 117, 173
Bhūmi Sūkta 11, 22, 149
Bhutan 6, 12, 52, 68, 82–5, 167
Bhutanese Forest Act *see* Forest Act biosphere 5
biotechnology 122, 171
Bishnois 4, 42, 56
BJP *see* Bharatiya Janata Party
bodhisattva 77,78, 90, 93, 167; self-negation of 12, 73, 79, 85
Bön religion 76, 79, 167
Bose, Jagadish Chandra 50, 56; influence of 11, 51, 63, 66, 166; philosophy of 43–5, 49, 169; research 34, 38, 42, 170

Index

Bose, Satyendra Nath 42
brahmacarya 27, 48
Brahman 24, 25, 26, 30, 38, 163
brahmin (*brāhmaṇa*) 23
Brahmo-Samaj 37, 42
Brandis, Dietrich 36
Bṛhad-āraṇyaka Upanishad *see*
  Upanishad(s)
Bṛhat Saṁhitā 64
British Antarctic Survey 121
Buddha *see* Siddhārtha
Buddhadāsa 85, 91, 92–6, 107, 109, 168
Buddhism *passim*, 6, 62, 63, 69–85, 152, 173; Chinese 74–5; Mahāyāna *passim*, 7, 12, 69, 70, 72–4, 76, 84–5, 90, 167; Nepalese 75–6; Theravāda *passim*, 7, 12, 77, 84, 90; Zen 93
Bunnag, Jane 9
Burma 76–7, 78, 82, 133

cakravartin 12, 72, 78, 90, 164
Callicott, J. Baird 8
Cambodia 77, 78, 107
Candragupta 27, 71
Carson, Rachel 1
cedar *see* deodar
Centre for Energy and Environment 14, 121, 138, 159
CFCs *see* chlorofluorocarbons
Chāndogya Upanishad *see*
  Upanishad(s)
Chandrasekhar 5, 43
Chapman, J. L. 5
Chapple, Christopher K. 6, 8, 71
China 2–3, 122, 131, 132, 133
Chinese Buddhism *see* Buddhism
Chipko 4, 14, 135, 140, 171, 174; emergence of 11, 12, 51, 58–61; Gandhian features 167; *see also* Bahuguna, Sunderlal; Bhatt, Chandi Prasad
chir (pine) 52, 54, 57, 59, 66, 169
chlorofluorocarbons (CFCs) 121
Christianity 6, 123, 164, 166, 169, 175; at Earth Summit 139, 147–8; influence on Gandhi 46; in Ladakh 79, 81; in Nagaland 134; relationship with secular state 110; responses to secularization 39–40; use of notion of jubilee 10, 162; Syrian Christians 6; *see also* Kocherry, Thomas; Roman Catholicism
Chukha hydro-electric power station 83
chula 65

Chulalongkorn 91
climate change *see* global warming
Collin, P. H. 5
Communist Party of India 112; *see also* Marxism
Connelly, James 6
consequentialist ethics 153–4, 175
consumerism 2, 84, 120, 126, 152, 172
Convention on Biodiversity 121
Coward, Harold 162

Dalai Lama 76, 79, 80, 81
dalit 112
darśana 26
Darwin, Charles 40–2, 49; *see also* evolution
Davidar, E. R. C. 147
deforestation *passim*, 4, 5, 35–6, 40, 88, 150
Delhi University 5, 34
deodar (cedar) 35, 52, 54
devarāj 78, 90
Dewan, Neelam 155
dhamma see dharma
dhammayuttika monks 91, 97
dhandak 56, 57, 60, 66
dharma, *dhamma* 27, 33, 37, 93, 105, 125; wheel of 12, 72, 84, 111
dharmarāja 164
Didayi tribal group 18
DMK *see* Dravida Munnetra Kazhagam
Dravida Munnetra Kazhagam (DMK) 112
Drèze, Jean 12, 110, 116, 126, 131–4, 174; *see also* Sen, Armartya K.
dukkha 69–70
Dumont, L. 9
dvaita 38
Dwivedi, O. P. 3

Earth Summit 3, 4, 5, 68, 83, 121, 137, 138, 139, 157, 169, 171, 172, 173, 175; religious contributions 13–14, 45, 118; northern bias 122, 135
ecology *passim*, 4–6, 78, 126, 154, 174; modes 17–18
Einstein, Albert 5, 41, 42, 49
Eliade, Mircea 9
Emerson, Ralph W. 1
energy *see* Centre for Energy and Environment; Tata Energy Research Institute
Engel, J. R. 5

Environment and Forests, Ministry of 120–1
Environment Protection Act 122
eucalyptus 58, 89, 118, 127
evolution 38; *see also* Darwin, Charles

famines 116
Fernandez, Walter 13
fishing rights 60; *see also* Kocherry, Thomas
Flood, Gavin 21
Forest Act(s) 122; Bhutanese 83; Indian Forest Act(s) 35, 54, 57, 122
Forster, E. M. 161
Four Noble Truths 69, 84, 164
Friends of the Earth 172
fundamentalism 165, 169

Gadgil, Madhav 12, 35, 110, 135, 152; early history 19, 29; environment-led analysis 126–31, 174; environmental debate 54; resource use 117, 118; 'scientific forestry' 55; *see also* Guha, Ramachandra; 'scientific forestry'
Gandhi, Indira: meeting with Sunderlal Bahuguna 59, 61; premiership 113–14, 116, 117, 124, 134, 149; and secular state 110; speech at Stockholm conference 4, 7, 119, 120
Gandhi, M. K. 11, 40, 51, 60, 65, 171; agrarian society 117, 127, 152, 173; ethics 45–50; influence on Chipko movement 61, 62, 63, 66, 67, 174; influence on Earth Summit 14; non-violence 153; notion of *sarvodaya* 34, 166; and *Ram Rajya* 111; salt march 57, 61
Gandhi, Rajiv 114, 115, 124
Gaṇeśa 28
gārhasthya 27
global warming 1–2, 4, 121, 159–60, 173
Godse, Nathuram 123; *see also* Hindu Right
Gokhale, G. K. 46
Golwalkar, M. S. 123, 125; *see also* Hindu Right
Gottlieb, Roger S. 8
Greenpeace 148, 172
Guha, Ramachandra 12, 35, 56, 61, 110, 135, 152, ; early history 19, 29; environment-led analysis 126–31, 174; environmental debate 54; resource use 117, 118, 'scientific forestry' 55; *see also* 'scientific forestry'

Haeckl, Ernst 5
Hamilton, Lawrence 5, 6
harijan 112
Harvard Center for the Study of World Religions 8
haṭha yoga 38
Hazare, Anna 11, 63–7, 51, 167
Hedgewar, K. B. 123; *see also* Hindu Right
Himalayas *passim*, 52–4
Hindu: Hindu tradition *passim*; definition of 7
Hindu Mahasabha 123; *see also* Hindu Right
Hindu Right 12, 110, 122–6, 164
hindutva 123, 125; *see also* Hindu Right
hydro-electricity *see* Chukha hydro-electric power station
hydrosphere 5

ICOREC *see* International Consultancy on Religion, Education and Culture
Indian Forest Act(s) *see* Forest Act(s)
Indonesia 76–7
Indra 22–3, 32, 163
Indus Valley civilization 20–1, 32, 163
International Consultancy on Religion, Education and Culture (ICOREC) 173
Iqbal, Mohammed 39; *see also* Islam
Islam 6, 113, 123, 166,175; alienation of Indian Muslims 111, 114, 122; Babri *masjid* incident 124; and conservation 168; influence on education 133; in Ladakh 78–9, 81; responses to secularization 39–40; World Muslim Congress 137
Īśvara 25, 28, 38

jagat 25, 39
Jainism 8, 46, 48, 70, 156
Jalal, Ayesha 115
James, George A. 3
Jana Sangh 112, 114, 124; *see also* Hindu Right
Janata Dal 112, 114
Janmabhoodi Programme 145, 147
Jātaka 71, 84, 92
jīva 25, 38
jubilee 10, 182: *see also* Christianity
Judaism 6, 10, 162

Kabilsingh, Chatsumarn 103, 104, 105
Kallayano, Phra Payom 95, 100, 101

karma *passim*, 26, 30, 31
karma-*yoga* 39, 40, 46, 48, 50, 166
Kauṭilya 27, 72; *see also* Arthaśāstra
Keyes, Charles F. 104
Khan, Sayyid Ahmad 39; *see also* Islam
Khāṇḍava forest 11, 16, 32–3, 148, 151, 154, 163
khandha 69
Khuttajitto, Phra Prajak 98, 108
Kinsley, David 8, 167; *see also padayātrā*
Kittiwuddho, Phra 97, 100, 101
Klausner, William J. 97
Kocherry, Thomas 147–8, 169, 171; *see also* fishing rights; Christianity; Roman Catholicism
Kosambi, Damodar 20, 22, 163
Krishna 11, 16, 32, 33, 148, 149, 151, 153–4, 163
Krishnan, K. S. 43
kṣatriya 23, 69

Ladakh 6, 12, 78–81, 82, 167; *see also* Norbert-Hodge, Helena
Laos 77, 78, 107, 160
Ling, Trevor 74, 165
liṅga śarīra 26
Lipner, Julius 25, 26, 28
lithosphere 5
Low, Nicholas 6

Madhva 24
Mādhyamika 74, 93
mae chii 98, 103–9, 161, 168; social roles 106
Mahābhārata 16, 25, 27, 163
Mahalanobis, P. C. 43
maha nikai 91, 97,
Mahāsāṅghika 73
Mahāyāna Buddhism *see* Buddhism
Malaysia 7, 77, 122, 168
Manu, Laws of 27
Marriott, M. 9
Marxism 11, 59, 61–2, 66, 112, 124, 128–9, 131, 171, 174
Mary, Avelin 147
māyā 39
medicinal plants (*samun prai*) 86, 97, 98, 145, 168, 176–80
Mehta, M. C. 150–2, 163, 175
methodology 8–10; *see also* Tambiah, Stanley J.
Ministry of Environment and Forests *see* Environment and Forests

Mongkut 88, 91, 92, 97, 107, 108, 165
Montreal Protocol 121; *see also* ozone layer depletion
multinationals 2
Munda tribal group 19
Muṇḍakya Upanishad *see* Upanishad(s)
Murcott, Susan 104

Naess, Arne 5
Nagaland 133–4; *see also* Christianity
Nāgārjuna 72–3, 74, 75
Nandī 29
Narasimha Rao, P. V. 115
Narayan, Jayaprakash 113, 114
Narmada Dam 83, 127, 136, 141, 144
Nehru, Jawaharlal 37, 58, 111, 113, 116, 117, 173
Nelson, Lance E. 8
Nepal 82, 137, 141
Nepalese Buddhism *see* Buddhism
New England Transcendentalists 1
nibbāna see nirvāṇa
nirvāṇa, nibbāna 6, 12, 70, 72, 73, 79, 85, 93, 99, 102, 162, 168
niṣkāma karma (selfless action) 31, 39, 46, 49, 63, 93; *see also* Bhagavadgītā
Noble Eightfold Path 70, 84, 164
nomads 18–19
Norbert-Hodge, Helena 79–80, 81, 85; *see also* Ladakh
nuclear waste 2, 171
nuclear weapons 2
Nun Institute of Thailand 105
nuns, Mahāyāna 79, 80; Thai *see mae chii*
Nyāya 26

Other Backward Classes 112, 115
ozone layer depletion 4, 42, 121, 122, 138, 173

padayātrā (pilgrimage) 11, 52, 59, 60, 61, 63, 66, 67, 167; *see also* Bahuguna, Sunderlal
Padmasambhava (Guru Rimpoche) 69, 76, 79, 82
Pakistan 122, 137, 147
Palmer, Martin 5
pañca-śīla 70
panchayat 47, 53, 54, 57, 58, 111, 147, 152; women's 65
panentheism 29
Paranjpye, Vijay 156
Parekh, Bhikhu 48, 50

pariṇāma 38, 42
Pārvatī 28
Paśupati 29
paṭicca samuppāda 70, 84, 93, 168
Pāṭimokkha 70, 84, 90, 97, 98, 106, 108
Patkar, Medha 13, 136–7, 141
Patna Devi 53
Payutto, Phra Prayudh 95
Photirak, Phra 101
pine *see chir*
population growth 2–3, 120, 150, 155, 159
Pradhan, Om 69, 83; *see also* Bhutan
prajñā-pāramitā 73, 79
prakṛti 25
Prance, Ghillean T. 5
Project Tiger 122
Public Interest Litigation 151; *see also* M. C. Mehta
Purāṇas 25
Pūrva Mīmāṁsaka 25, 26

Rahula, Walpola 164
Rajan, M. G. 3
rakhi 59, 67
Ramaiah, Darepalli 146
Ramakrishna 38, 150
Ramakrishna Mission 38, 166
Raman, C. V. 43
Rāmānuja 24, 26
Rāmānujan, Srinivasa 42
Rāmāyaṇa 25, 27, 57, 61
Rao, Raja 161
Rashtriya Swayamsevak Sangh (RSS) 123, 124, 125, 126, 141; *see also* Hindu Right
Raturi, Ghanshyam 59
Reddy, Amulya K. N. 142, 143
Redfield, Robert 53
religion *passim*, especially 6–8
Ṛgveda 21, 23
Rigg, Jonathan 88
Rimpoche *see* Padmasambhava
Roman Catholicism 6, 169; Vatican 13; *see also* Kocherry,Thomas
Roy, M. N. 37
Roy, P. C. 41–2, 44
Roy, Ram Mohan 31, 37, 92, 162
RSS *see* Rashtriya Swayamsevak Sangh
Ruskin, John 46–7

Sachs, Wolfgang 6
sacred groves 155–7
'saffron brigade' 65, 161; *see also* Hindu Right; Rashtriya Swayamsevak Sangh

Saha, Megnad 42
St Stephen's College 34; *see also* Christianity
St Stephen's Hospital 110; *see also* Christianity
Śaiva Siddhānta 26
sal 17, 35, 52, 54, 69–70
Sambad Prabhakar 41
Sāṁhita(s) 21
Śaṁkara 24, 26, 161, 163
Sāṁkhya 26, 38, 42, 44
saṁnyāsa 27
saṁsāra 30, 73, 93
sanātana dharma 150
Sangh Parivar 125; *see also* Hindu Right
saṅgha passim, 72, 75, 78, 90, 91, 96, 97, 161, 165
Saṅgha Acts 91
Sarasvati, Dayananda 37, 38, 122, 164
Sarnath Lion Capital 111, 173
sarvodaya 34, 47, 48, 50, 58, 63, 67, 166; *see also* Ariyaratne, A. T., Gandhi, M. K.; Sri Lanka
Śatapatha Brāhmaṇa 21–2; *see also* Agni
satellite imagery 135, 140, 171
satyāgraha 47, 48, 50, 166; *see also* Gandhi, M. K.
Savarkar, Veer D. 123; *see also* Hindu Right
'scientific forestry' 55, 56, 67
Seal, B. N. 42
secular state 110, 111–13, 173–4
secularization 36–7, 49, 92
Sen, Amartya K. 12, 110,148, 152; consequential ethics 153–5, 175; definition of development 116; development-led analysis 131–4, 174
Sen, Keshub Chandra 37
Sharma, P. D. 5
Shastri, Lal Bahadur 112, 113
Shiva, Vandana 169
Siddhārtha, the Buddha (Gotama) 69–72
Sikhs 6
Singh, Karan 11, 29–32, 148–50, 163, 164, 175
Singh, Tavleen 136
Śiva 21, 28, 29, 38, 77, 150
Sivaraksa, Sulak 5, 99, 100
Smart, Ninian 6, 165
smṛti 25
Sri Lanka 6, 12, 72, 77, 78, 133, 137; early Buddhism 74–5; work of A. T. Ariyaratne 166
Srinivas, M. N. 37

210 Index

śruti 25, 163
Sudarshan, Hanumappa Reddy 146
Sudarshan, K. S. 126; see also Hindu Right
śūdra 23, 112
Sumatra 76
śūnyatā 72–4, 79, 85, 93, 168
sustainable development 3
Sutta 70
Śvetāśvatara Upanishad see Upanishad(s)
swadeshi 48, 50, 166
Swadhyay movement 144
Swaminarayan sect 144
swaraj 48, 50, 166
Swearer, Donald 94–5, 167
Sweemer, Cécile de 160
Syrian Christians see Christianity

Tagore, Debendranath 37, 40
Tambiah, Stanley J.: galactic polity 90; methodology 9, 32, 35, 45, 49, 52,162, 164; neotraditionalism 92; Thai monasticism 95
tantra 38, 68, 75, 76, 77, 78, 83, 167
Tanwar, Sukhbir 145
Taoism 75
Tata Energy Research Institute 14, 121, 38, 141
Tattvabodhini Patrika 41
Tehri Dam 140, 141
Thailand chapter 6; 12, 77, 78, 133, 168, 172
Thakker, Anuben 145
Thapar, Valmik 147
Theravāda Buddhism see Buddhism
Therigatha 104
Thomas, M. M. 7, 40
Tibetan Buddhism 74, 76, 79, 82
Tovivich, Phichai 103
trikāya 74, 102
Tripiṭaka 105

United Nations: Earth Summit (1992) 13–14, 118, 135, 157; Nairobi Conference on New and Renewable Energy Sources (1981) 120; Stockholm Conference (1972) 4, 7, 88, 119; population policy 3
Upadhyaya, Deen Dayal 124, 125, 126; see also Hindu Right

Upanishad(s) 14, 22, 24–7, 38; Bṛhad-āraṇyaka Upanishad 24; Chāndogya Upanishad 24, 26, 30; Muṇḍakya Upanishad 30; Śvetāśvatara Upanishad 25
utar 57
Uttara Mīmāṁsaka 26

Vaiśeṣika 26
Vaishnavism (Vaiṣṇavism) 38, 45
vaiśya 23
Vajpayee, Atal Behari 111, 114
vajra 22, 32, 163
Vajrayāna 74, 76, 79, 85
vana mahotsava 149, 175
Varāhamihira 64
varṇa 23, 27
varṇāśrama dharma 27, 163
Vatican see Roman Catholicism
Vedānta 32, 163, 164; and Jagadish Chandra Bose 34, 38, 167; schools of 25, 26, 30; and science 41, 49
Vijñānavādin (Yogācārin) 74
Vinaya 70, 75, 91, 97
viññāna 69, 162
Vishnu (Viṣṇu) 33, 77, 127
Vishwa Hindu Parishad 123, 125, 141; see also Hindu Right
viśiṣṭādvaita 38
Vivekananda 34, 45; communality among life forms 11, 46; ethics 38–9; influence on M. K. Gandhi 48, 50, 166; influence on Anna Hazare 51, 64, 66

Wasi, Prawase 96, 98, 108, 168
Wat Phra Thammakaay movement 100, 102
Wildlife Protection Act 122
winyān 162
women's literacy 132
women's rights 3, 27, 36, 60, 103–7, 150
World Council of Churches 10
World Wide Fund for Nature 145, 172

yajña 21, 25
Yajurveda 21, 23–4
Yoga 8, 26

Zen Buddhism see Buddhism
zamindari 112, 115